AF477168

INTRODUCTION TO
NANOSCIENCE AND NANOTECHNOLOGY

P. Venugopal Reddy Ph.D., FTAS, FAPAS

Professor of Physics & Dean Exams.
Vidya Jyothi Institute of Technology,
Hyderabad- 500 075

M. Lakshmi
Assistant Professor,
Vidya Jyothi Institute of Technology, Hyderabad.

BSP BS Publications
An imprint of **BSP Books Pvt., Ltd.**
4-4-309/316, Giriraj Lane,
Sultan Bazar, Hyderabad - 500 095.

Introduction to Nanoscience and Nanotechnology

by *P. Venugopal Reddy and M. Lakshmi*

© 2023, *by Publisher,* All rights reserved.

Published by:

BSP **BS Publications**

An Imprint of BSP Books Pvt., Ltd.
4-4-309/316, Giriraj Lane,
Sultan Bazar, Hyderabad - 500 095.
Phone: 040 - 23445688
e-mail: info@bspbooks.net
www.bspbooks.net

ISBN: 978-93-95038-11-9 (Hardback)

Preface

Although several books on nanoscience and nanotechnology are available in the market, most of them deal with the research level areas of some field or other, while a few of them deal with fundamentals or basics of the subject. In fact, most of the books available in the market are not suitable for undergraduate students. In order to fill the gaps, and to serve the needs of undergraduate students sincere efforts have been made by the authors of the book. This book also serves the students who are studying the Nanoscience & Nanotechnology course for the first time at undergraduate or even at post graduate level.

Nanotechnology is a multidisciplinary subject representing the convergence of various subjects. Convergence has been manifested in areas such as Physics, Chemistry, mathematics, materials science, electronics, medicine, informatics, biotechnology, genetic engineering, etc. This convergence, has resulted in the disappearance of boundaries among the subjects leading to the creation of new disciplines. For example, nanobiotechnology is the result of convergence of nanotechnology and biotechnology. This implies that a student of this new area should be in a position to understand the confluence of various subject boundaries. In such a situation, how the pedagogy adopts to the rapidly evolving and smearing boundary disciplines? This is the central issue while teaching NS&NT. Moreover, this course will be useful to the students who wish to pursue their higher studies in nanotechnology. In fact, the study and development of the subject at nanoscale is possible because of the discovery of several instruments allowing them to see, measure and manipulate the nanoparticles.

In fact, for the people who are practicing NS&NT may find it difficult in structuring the curriculum for the subject. A unified curriculum dealing with every aspect of the subject is indeed a difficult task, especially when the subject boundaries are rigid. Although most of our traditional subjects do not allow any serious modifications, every effort is made to cover the subject to a large extent. As Nanoscience & technology is in an infant stage, expectations are very high and it is nearly impossible to cover everything in a book like this. Despite these limitations, the authors hope that the book meets most of the expectations.

Prof. P. Venugopal Reddy

Dr. M. Lakshmi

Summary and Acknowledgements

The book and it contains nine chapters. **Chapter 1** is an introductory chapter and begins with the definitions of nanoscale, nanoscience, nanomaterials, nanotechnology etc., This chapter illustrates the general introduction, historical aspects, origin and general applications of nanotechnology. Nanostructures play a significant role in the advancement of scientific and engineering technologies at the nanoscale. Therefore, all these aspects are included in the introduction chapter.

Chapter 2: An indepth knowledge of various types of nanostructures is essential for understanding the subject. The concepts of subjects are very much difficult to understand and as such every effort is made to explain the basics and later important aspects of the subject. The chapter also deals with variety of biological structures such as DNA, gecko, dazzling colours of butterfly, peacock feathers etc. All the details regarding the influence of nanoscience and technology on various types of living organism are explained in this book.

Chapter 3: Provides detailed information about the classification of nanomaterials. Based on the quantum confinement, nanomaterials may be classified as a) Zero dimensional b) One dimensional c) Two dimensional and d) Three dimensional nanomaterials. Details regarding the surface to volume ratio along with its importance in Nanomaterials is given in this chapter.

Chapter 4: Deals with the synthesis of nanomaterials by various methods. The synthesis of nanomaterials using top-down and bottom-up methods have been presented in this chapter. Top-down methods such as Solid State Reaction process, hydrothermal process, Solvothermal process, Nanolithographic process have been discussed in detail. Similarly, the a detailed discussion on the bottom Up methods including Sol- Gel method, Chemical Vapour Deposition process, Physical Vapour Deposition Process, Pulsed Laser Deposition technique, Sputtering Technique are given in this chapter. Chapter-5 focuses on different characterization techniques which are used in analysing nanoscale materials. They include X- Ray Diffraction (XRD), Fourier Transform Infrared spectroscopy (FTIR), Scanning Electron Microscope (SEM), Transmission Electron Microscope (TEM), Scanning Tunnelling Microscope (STM), Atomic Force Microscope (AFM). A detailed discussion on all these topics is given in this chapter.

Chapter 6: Gives a complete description of different types of properties of several of nanomaterials. They include Physical, Chemical, Optical, Magnetic, Thermal, Mechanical, and Electrical and Electronic properties of nanomaterials. Some of the topics in this chapter include a detailed discussion on Moore's law and its present status are given. Synthesis and applications of nanofluids have also been included in this chapter.

Chapter 7: Nanoscience & nanotechnology plays a dominant role in number of modern technologies. The unique properties at the nanoscale make nanomaterials to have several applications in various fields. They include electronics, computers, textiles, catalysis, paintings, coatings, personal care products, energy savings, alternative energy supplies, environmental protection, agriculture applications, medical uses etc., All the details are given in this chapter.

Chapter 8: A detailed discussion on nanoscale devices with applications in photovoltaics, medical diagnostics and electronic devices are given in this chapter. In fact, the outstanding performance of nanomaterials made them useful in developing variety of nanoscale devices. The fascinating and often incomparable applications opened up new and sometimes unexpected devices. Today, the widespread applications of these devices range from drugs, bio- and infrared-sensors, spintronic devices, data storage media, magnetic read heads for computer hard disks, single-electron devices, microwave electronic devices and many more.

Chapter 9: Deals with some of the important concerns and challenges of nanotechnology. Potential risks include environmental, health, and safety issues, transitional effects such as displacement of traditional industries as the products of nanotechnology become dominant. These may be particularly important if negative effects of nanotechnology are overlooked.

Acknowledgements

The authors acknowledge the help and encouragement given by Dr. P. Rajeshwar Reddy, Secretary and Correspondent of Vidya Jyothi Institute of Technology.

Contents

Chapter 1
Basics of Nanoscience

Chapter 2
Nanostructures

Chapter 3

Classification & Surface to Volume Ratio in Nanomaterials

Chapter 4

Synthesis of Nanomaterials

Chapter 5

Characterization Techniques

Chapter 6

Properties of Nanomaterials

Chapter 7

Applications of Nanomaterials

Chapter 8

Nanoscale Devices

Chapter 9
Concerns and Challenges of Nanotechnology

Chapter 1

Basics of Nanoscience

1.0 Basic Definitions

1.1 Definition of Nano

Nano comes from the Greek word "nanos," meaning "dwarf", but nano is infinitely smaller than a dwarf. Nano means 10^{-9}. Nanometer is one billionth of a meter (i.e. 1nm= 10^{-9} m). Some examples are given in the following figure to understand how small the nano dimension is 1centimeter = 10^{-2}m, 1 millimeter = 10^{-3}m, 1 micrometer = 10^{-6}m and 1 nanometer = 10^{-9}m.

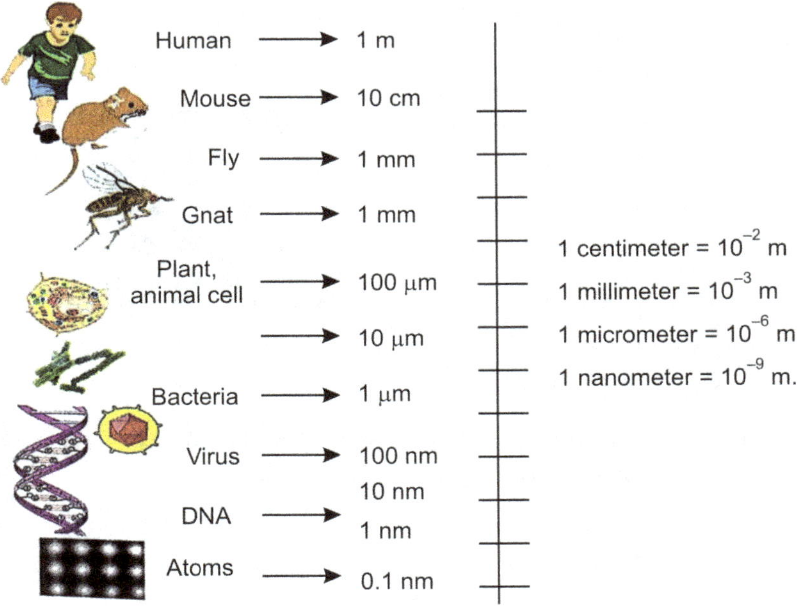

Fig. 1.1 Sequence of images showing various levels of scale

1.2   Nanoscale

Nanometer is the scale used to measure objects in the nanoworld. In general, the scale ranges from **1nm to 100nm** is known as **Nanoscale**. It is appropriate and relevant to make a mention of the diameter of an atom. As the size of an atom is less than even 1 nanometer, we discuss the next scale on the lower side and is Angstroms (**Å**) scale. We are aware that atoms are extremely small and the diameter of the atoms is measured in Angstroms (**Å**). Thus, 1 Angstrom (**Å**) = 0.1 nm. For example the diameter of a hydrogen atom is 0.1 nm. If one lines up 10 hydrogen-atoms next to each other in a row, the resulting length would be approximately 1 nm.

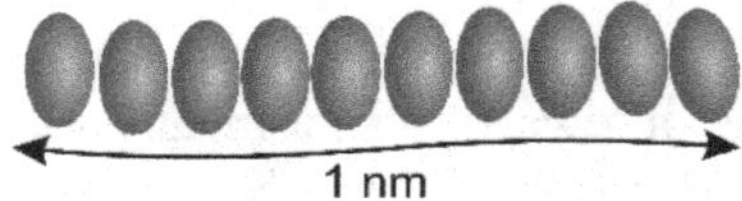

Fig. 1.2 Ten hydrogen atoms in a row make 1 nm

To put this scale in perspective, "Trying to find a nanoparticle on the surface of a soccer ball is just as difficult as finding a soccer ball on the surface of the Earth".

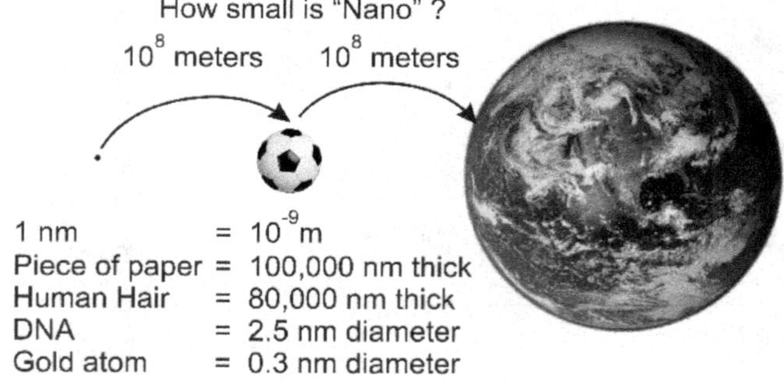

Fig. 1.3 Nanoparticle compared in size with a soccer ball, and soccer ball is compared with Earth (figure by Michael Hochella https://serc.carleton.edu/details/images/180059.html)

1.3   Nanomaterials

The material whose crystallite size lies in the nano range (1nm to 100nm) is called **nanomaterial**. The properties of the material at the nano scale (Nanomaterial) are totally different from those of the same material at the higher scale (Bulk material). It means that most of elements of the periodic table exhibit totally different behavior when they are in the nanoscale. For example, although the gold is chemically inert in its bulk state, in the nano state, however, exhibits excellent catalytic activity. Similarly, bulk silver is non-toxic, whereas nano-silver is capable of killing viruses. Further, Titanium dioxide, a white pigment used in sun screens, becomes transparent in the nanoscale, while aluminum oxide, commonly used for teeth fillings, becomes explosive in the nanoscale. Finally, one may conclude that Nanomaterials exhibit improved properties such as strength, hardness, ductility (in brittle materials), wear - resistance, corrosion – resistance, erosion resistance, improved chemical activity etc., Moreover, as classical mechanics fails to explain the unusual behavior (movement, energy, etc) of the nanomaterials, quantum mechanics, however, is successful in doing so.

1.4   Nanoscience

The science dealing with all types of nano materials including nanoparticles, nanostructures and their unique properties is defined as Nano science.

Nanoscience is the cutting edge of science and is highly multidisciplinary. It draws from diverse fields such as physics, chemistry, materials science, microelectronics, biochemistry, Engineering and biotechnology etc.

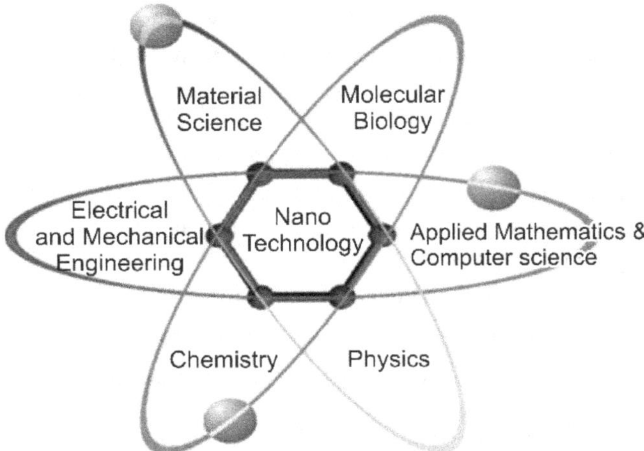

Fig. 1.4 Nanoscience is a multidisciplinary subject

1.5   Nanotechnology

The technology involving the manipulation of nano materials to build variety of structures is defined as nano technology. Although Nano technology is still at its infant stages, the nano products are having far reaching applications and are going to change the lives of mankind. **Nanotechnology** helps in the design, characterization, production and application of structures, devices and systems by controlling size and shape at the nanometer scale (1–100 nm). Nanotechnology is very much useful in creating these novel materials and devices with enhanced properties and potential.

In recent years nanotechnology has become highly popular. This is partly because of the interest in miniaturization of devices. Scientists have also been amazed due to the extraordinary properties of these novel materials. In fact, these materials open a new avenue in the field of material science due to their novel properties and wide range of applications in various fields such as Electronics and communications, Textile industry, Automobiles, Sensors, Cosmetics and Medicine etc.

1.6 General Introduction

Nanomaterials open a new avenue in the field of material science. The nanomaterials have been the focus of intense research recently because of their novel properties at the nanoscale. Nanotechnology is a field of applied science concerned with the control of matter at dimensions of roughly 1 to 100 nanometers (nm). Nanotechnology is a new and expanding technology, its main applications are the development of innovative methods to fabricate new products, to formulate new chemicals, materials and efficient use of raw materials. Nanotechnology is also meant for replacing the recent era of equipment with better performance ones. This will help in the reduction of consumption of energy and material and lowered damage to the environment, as well contributing to the environmental remediation [1]. The widest use of nanomaterials has been as tumor-targeted drug delivery systems, contrast enhancers for magnetic resonance imaging, biosensors and biomedicine, etc.

Nanomaterials are already in commercial use. The range of commercial products available today is very broad, including sunscreens and cosmetics, paints, UV-blocking coatings, self - cleaning windows, fuel cells, batteries, fuel additives, lubricants, catalysts, stain-resistant and wrinkle-free textiles, transistors, lasers and lighting, integrated circuitry, sports equipment, bicycles and automobiles, medical implants, water purifying agents, disinfectants, food additives etc.

In order to design nanomaterials and nanodevices for the next generation, it is requisite to understand their intrinsic features. Several researchers are working throughout the world towards the development of nanomaterials that are intended to perform more complex and efficient tasks. Thus, the development of completely new technologies and nanomaterials with desirable functional properties may lead to a generation of new products that will enhance the quality of the living environment in the near future.

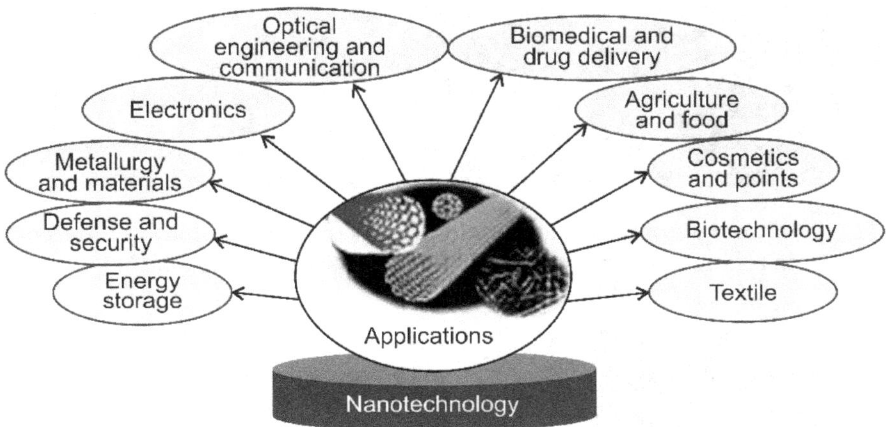

Fig. 1.5 Applications of Nanomaterials [1]

1.7 History of Nanotechnology

The origin of the term 'Nano' is traced to a Greek root which means dwarf (very small). On the length scale, nano is one billionth of a meter. The word Nanotechnology is relatively new, but the existence of nanostructures and nanomaterials is not new. Such structures have existed on Earth as long as life itself. Nanomaterials have been used by human beings for centuries without knowing the dimensions of the materials they used. The first known use dates back to 4th century A.D. Roman glass makers have prepared glasses containing nanosized metals. The Lycurgus cup is an outstanding example of this. It appears green in the reflected light and deep red in transmitted light. This cup is made from soda lime glass containing silver and gold nanoparticles inside it. The glass is found to contain 70 nm particles when seen through the transmission electron microscope (TEM). Nanotechnology is easily evident in various old churches. The beautiful colors on window glasses of medieval churches are due to the presence of metal nanoparticles in the glass materials. These vivid colors were controlled by the size and the shape of the nanoparticles of gold and silver.

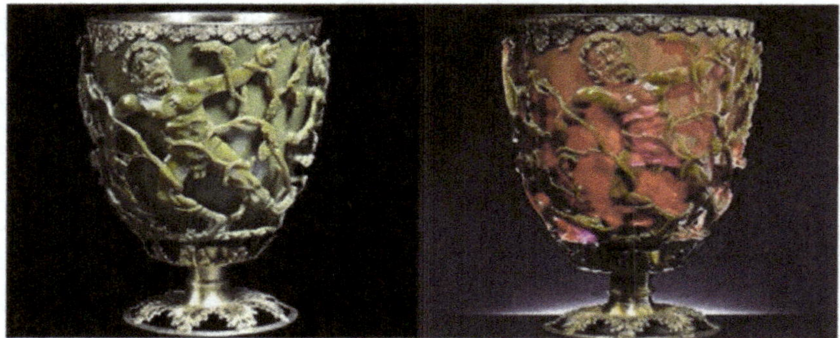

Fig. 1.6 Photographs of the famous Lycurgus cup which displays a different color in reflected and transmitted light. [Courtesy of the British Museum, London]

- Red Ag (~100 nm, Triangle)

- Yellow, Au (~ 100 nm, spheres)

- Green: Au (~ 50 nm, Spheres)

- Light blue Ag (~90 nm, spheres)

- Blue Ag (~40 nm, Spheres)

Fig. 1.7 Rosacenord stained glass in the Cathédrale Notre-Dame de Chartres (France),color changes depend on the size and shape of gold and silver nanoparticle. [2]

The Damascus swords were prepared in Damascus, the capital city of Syria, during 900 AD to 1750 AD. Damascus steel is a kind of alloy. With this alloy sharp, flexible and hard swords were manufactured. They were famous for their strength and sharpness. It is said that the blade of Damascus sword is very strong and even when it falls on the ground retains its original sharpness. In fact, this sword was used for cutting not only wood but also metal or even stone.

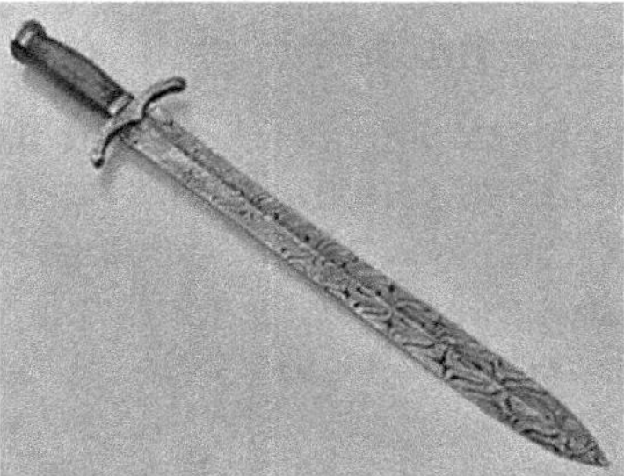

Fig. 1.8 Damascus sword [3]

What gave the sword its uniqueness? The secret of the sword remained unknown until recently. Very recently, high resolution electron microscopy of blades used in the Damascus sword showed that the blade contains carbon nanotubes (CNTs) [3]. Therefore, now people also strongly believe that the existence of carbon nanotubes (CNTs) in the material is the main reason for the high strength of these steels.

A medical Doctor / practioner Francisci Antonii published the first book on colloidal gold. It contains a detailed description of the formation of colloidal gold and covers its various medical applications. The book noted the presence of soluble gold around fifth or fourth century B.C. in Egypt and China.

The field of black and white photography was started in the 18th century. The photographic film is an emulsion consisting of silver nanoparticles in gelatin.

The first preparation of nanoparticles in the laboratory was carried out by Michael Faraday as early as 1857 [4]. He recognized that the colour was due to the small size of the colloids and published a paper explaining how metal particles affect the colour of church windows. The gold particles prepared by Faraday are still preserved in the Royal Institution of London.

Fig. 1.9 A display of Michael Faraday's colloidal solutions of gold at the Royal Institution [4]

1.8   Origin of Nanotechnology

Today's extraordinary developments in the fields of nanoscience and nanotechnology are actually based on the ideas of some legendary scientists of last century. Science and Technology always moves forward and newer, developed devices replace the older one. In the last decade, new dimensions of modern research in the field of "nanoscale science and technology" have been emerged. The ideas and concepts behind nanoscience and nanotechnology started with a lecture entitled "There is Plenty of Room at the Bottom" by Nobel Prize winner, famous Physicist Richard Feynman, in the annual general body meeting of the American Physical Society at California Institute of Technology on December 29th in 1959. In his lecture, he has speculated on the possibility and potential of nanosized materials. Many of the speculations of Faynman have become reality (i.e. nano silicon chips in nanoelectronics, nanostructures in nanomachines, biological nanoparticles in medicine). Therefore, people consider him as the father of Nano Science & Technology

Fig. 1.10 Richard Feynman

The term "Nano-technology" was first used by Norio Taniguchi in 1974 to explain the new semiconductor technology of thin-film deposition that deal with the controlling the size of thin film of very small in size of the order of nanometers, though it was not widely known at that point of time. In 1981, the scanning tunneling microscope (STM) was invented by Gerd Bining and Heinrich Rohrer at IBM research laboratory. Another nano technology breakthrough occurred in 1985 by a team led by Richard Smalley and Harry Kroto in discovering a new compound, C60, a carbon nano particle with a shape of a soccer ball (Bucky ball). In 1991, Iijima made carbon nanotubes (CNT). Advancements in microscopy technology have made it possible to visualize several nano structured materials and have largely dictated the development of nanotechnology.

Inspired by Feynman's concepts, K. Eric Drexler used the term "Nanotechnology" in his 1986 book Engines of Creation: The Coming Era of Nanotechnology. In his work, he used the term "Nanotechnology" in 1986 to give an idea about building machines at molecular level even far smaller than a cell. K. Eric Drexler spent number of years in analyzing these incredible devices and finally published a famous book in 1990 "Machinery Nanosystems: Molecular, Manufacturing, and Computation". Feynman and K. Eric Drexler certainly popularized nanotechnology. In 1994, stable gold nanoparticles were prepared in the form of a solution. In 1996, scientists at IBM laboratory succeeded in moving and accurately positioning individual molecules at room temperature. In 2004, Andre Geimand Konstantin Novoselov discovered the two-dimensional nano material (Graphene), for which they were awarded with a Nobel Prize in 2010. From 2005 onwards a series of R& D activities helped the new branch to develop into a full-fledged multi-disciplinary subject.

References

1. Mahmoud Nasrollahzadeh, S. Mohammad Sajadi, Mohaddeseh Sajjadi, Zahra Issaabadi, Chapter 4 - Applications of Nanotechnology in Daily Life, Interface Science and TechnologyVolume 28, 2019, Pages 113-143

2. Jin , R. , Cao , Y. , Mirkin , C.A. , Kelly , K.L. , Schatz , G.C. , and Zheng , J.G., Photoinduced Conversion of Silver Nanospheres to Nanoprisms, Science Volume 294 , 2001, Pages 1901– 1903.

3. M Reibold, Peter Paufler, Alexander A. Levin, W Kochmann, N Pätzke, Dirk Carl Meyer Materials - Carbon nanotubes in an ancient Damascus sabre, Nature, Volume 444, 2006, page286.

4. Michael Faraday, The Bakerian Lecture: Experimental Relations of Gold (and Other Metals) to Light, Philosophical Transactions- The Royal Society London, Volume 147, 1857, Pages 145-181.

Chapter 2

Nanostructures

Nanostructure is an object that has at least one of the three dimensions is equal to or smaller than 100 nanometers. Based on the orientations, morphologies and their characteristics, nanostructures may take the shape of nanowires, nanofibers, nanotubes, nanobelts, nanofluids, nanoribbons, nanosprings, nanocapsules, nanosheets etc., Nanostructures constitute a very active area of research and development in the fields of nanotechnology and nanoscience. Nanostructures play a an important role in the development of science and engineering technologies. In recent years nanostructures have drawn much attention due to their unique characteristics at nano dimensions that influence the physical, chemical, optical, electrical, biological, etc., properties of the materials.

Nanostructured Materials (NsM) may be in or far away from thermodynamic equilibrium. The characteristics of NsM vary from those of single crystals or polycrystals and even glasses with a similar chemical composition. This variation is due to the reduced crystallite size as well as from the innumerable grain boundaries between adjacent crystallites at the nano dimensions.

The composition of nanostructures, such as monometallic, bimetallic, magnetic, metal oxide, semiconductor, hybrid, composite, etc., has been used frequently as the basis for their classification.

2.0 Manipulation of Atoms

It is possible to manipulate the atoms using various techniques so that varieties of nanostructures are possible. For example, Donald Eigler and Erhard Schweizer scientists from IBM laboratory were successful in manipulating with the xenon atoms to spell out the IBM logo as shown in Fig 2.1. A scanning tunneling microscope (STM) was operated to place 35 individual Xe-atoms (xenon atoms) on a chilled nickel crystal substrate to spell out the company acronym "IBM". It was the very first attempt atoms had been exactly arranged over the flat surface [1].

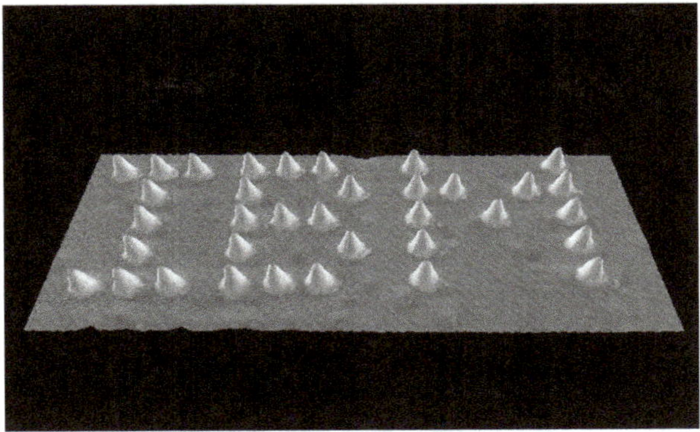

Fig. 2.1 35 Xenon atoms spelling out the IBM logo

2.1   Quantum Confinement (Corral)

The 48 iron atoms, adsorbed onto the copper surface, were moved into position with the tip of a low-temperature scanning tunneling microscope, to make a ring with a diameter of 143 Å. An STM image of a circular quantum corral constructed from 48 Fe atoms is shown in Fig. 2.2. The circular pattern in the center of the corral is the density distribution due to 3 nearly degenerate quantum states of the corral. It resembles a birth day cake [2].

Fig. 2.2 Quantum corrals on a Copper surface

2.2   Building of Nanostructures

Nanotechnology means building of variety of structures using atoms that too one atom at a time. For example, K. Eric Drexler designed a nanoscale planetary gear and connects an input shaft through a sun gear to an output shaft via a cluster of planet gears. The planet gears rotate between a ring gear and the sun gear on the internal surface of the casing.

Fig. 2.3 A nanoscale planetary gear

2.3   Nano Instrumentation

Scientific research on nanotechnology has led to materials science entering a new era known as "nanomaterial era". Over the past three decades, several microscopes and other instruments needed for analyzing and studying the properties of nanomaterials were developed. They include Atomic force microscopy (AFM), Scanning Tunneling Microscope (STM), Tunneling Electron Microscope (TEM) etc., These powerful nanotechnological imaging tools, has helped a lot in understanding the topography, elasticity, adhesion, friction, electrical properties, and magnetism of these materials. Observation of a few more types of nano structures by STM and AFM are shown in figures 2.4 and 2.5 below.

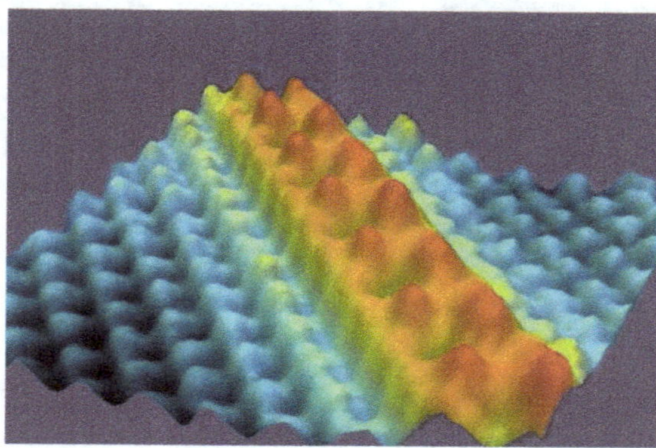

Fig. 2.4 STM image of chain of Cs atoms (red) on GaAs surface (blue) [3]

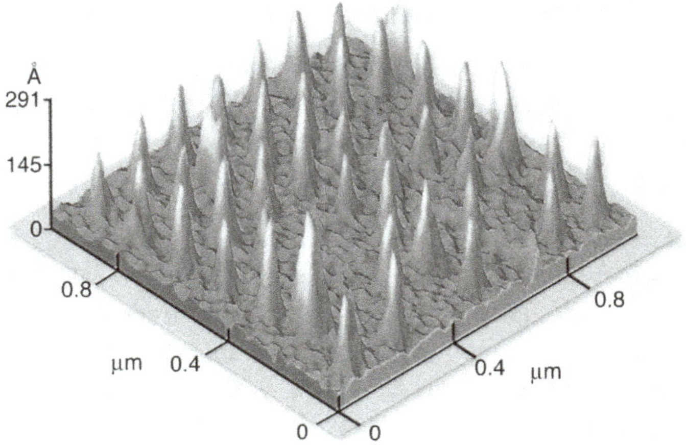

Fig. 2.5 AFM image of 20-nm Pt nano particles 100 nm apart

2.4   Observation of Magnetic Flux Lines of Nickel Nanoparticles

Another remarkable achievements of Nano science is the observation of magnetic fluxlines in a magnetic material. Normally, it is not possible to see the flux lines of a magnetic material. These figures (Fig. 2.6) will see while performing an experiment in magnetism – Magnetic lines of force.

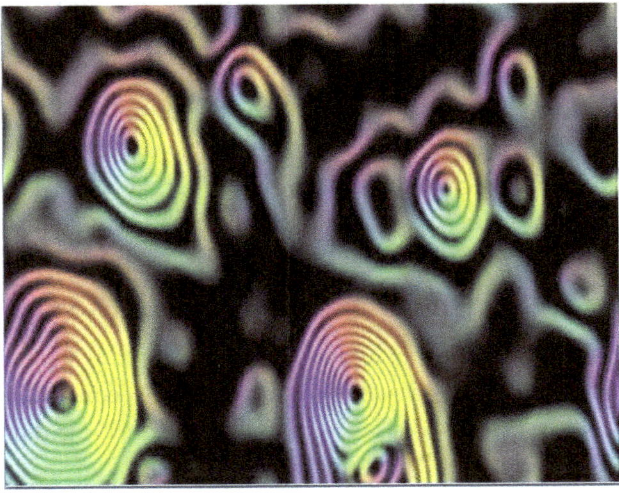

Fig. 2.6 Magnetic flux lines of nickel nanoparticles

2.5   Nanostructures in Nature

Nature is the great experienced and tested laboratory ever known to us and skilled in creating sophisticated materials, self-healing, capturing energy and storing information with excellent efficiency.

Naturally Produced Nanomaterials

Nanostructures are plenty in nature. In the universe, nanoparticles are distributed widely and are considered to be the building blocks in the planet formation processes. Indeed,several types of nanostructures are present in living organisms ranging from microorganisms, such as viruses, nanobacteria and nanobes, algae, fungi, yeast and bacterial spores to complex organisms such as plants, birds, insects, animals and humans. Latest and sophisticated equipment help us in better understanding of the naturally available of these organisms. In fact, a detailed knowledge of the nanostructures which are present in microorganisms is vital for identifying the further use of these organisms for various biomedical applications.

Several insects are having nanostructures and might have been created by an evolutionary process and are in a position to survive even in harsh living conditions because of their nanostructural constitution. Plant absorbs the nutrients from the water and soil for their growth which leads to the accumulation of these biomaterials in the nano structured-form. Similarly, humans also have organs primarily built with nanostructures like enzymes, bones, antibodies and other secretions that are highly useful for the appropriate function of humans. It is also to be noted that DNA or RNA are also nanostructures. These are essential for the formation of cells and the function of all living organisms. One may also note that nanostructures are the basic foundation for all types of living organisms. The following sections explain various types of nanostructures that are present in the biological world.

2.6   Gecko Feet Extraordinary ability of Gecko Feet

Gecko (a type of lizard) has an extraordinary ability to adhere to a surface. Gecko runsfrom one end of a wall to other in a few seconds. It even moves easily upside down across a ceiling of our houses. This typical property may be explained on the basis of the fact that millions of nanosize hairs present on each foot of gecko produces adhesive force and their combined adhesive force is 100s of times greater what is required for the gecko to hang to the ceiling. The hairs can readily bend to conform with the topography of a surface, which allowsgeckos to stick to even the flat surfaces such as glass. This discovery has launched several research groups into developing various products ranging from reusable adhesive tape to shoe soles for climbing robots.

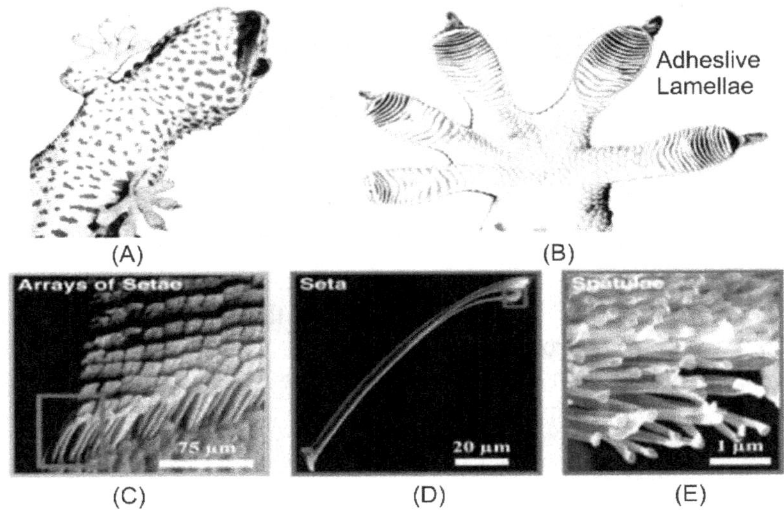

Fig. 2.7 Structural hierarchy of the gecko adhesive system.
(A) Macrostructure: ventral view of a gecko (B) Mesostructure: ventral view of the foot, (C) Microstructure: proximal portion of a single lamella, with individual setae in an array visible. (D and E) Nanostructure: single seta with branched structure at upper right, terminating in hundreds of spatular tips [4]

2.7   Dazzling Colors of Butterfly Wings

The spectacular colors of butterfly wings are not the product of pigmentation, but theirwings consist of many air-sandwiched layers of nano-scale structures [5]. When the light incidents on these layers, the innumerable reflections produce constructive interference patterns leading to dazzling colors. The nanostructures also contribute to the selection of colors by only reflecting a few selected wavelengths. Based on the size, shape, angle and spacing of the nanostructures, particular colors are selectively strengthened or cancelled out giving beautiful colors to the butterfly wings. The nano structure layers of butterfly wings inspire scientists to develop textiles by assembling nano particles into layer by layer using the bottom up method, so that the textile material exhibits different kind of colors in the indoor light and another set of colors in the outdoor light.

Fig. 2.8 Dazzling colors of butterfly wings

The butterfly wings not only exhibits dazzling colors but also it has a very good hydrophobic property. Because of the hydrophobic surface of butterfly water rolls from the surface of its wings taking all the dirt. Therefore butterfly wings are having very good self- cleaning property because of its super hydrophobic nature.

2.8   Lotus Effect

The water-repellent and self-cleaning property of lotus leaf is due to the presence of nanosized wax papillae on the surface of leaves [6]. When the rainwater falls on the lotus leaves surface, make a high contact angle and roll down, carrying dust particles with them, leaving the leaf surface clean and making the leaf surface hydrophobic or highly water repellent. This self-cleaning property of the lotus leaf surface is called the lotus effect. In fact, this property has opened possibilities of fabricating super hydrophobic surfaces for a variety of applications.

Fig. 2.9 Water drops on hydrophobic surface of Lotus leaf

Among the numerous applications of lotus effect are, sealing windshields, waterproofing phones, non-wettable rain wear and sails for boats, water repellent paints on furniture, paints for kitchen roofs and walls that make them soot-free, self-cleaning coatings on the cars, windows in high-rise buildings and glass for greenhouses avoiding their expensiveand cumbersome cleaning, non-sticky bottles, super hydrophobic coating based plastics, water and dirt repellant fibers for textiles etc. Other applications are sanitary products especially in toilets and bathrooms and motor vehicle windshields for decreasing sticking of dirt and more effortless cleaning.

2.9   Peacock Feathers

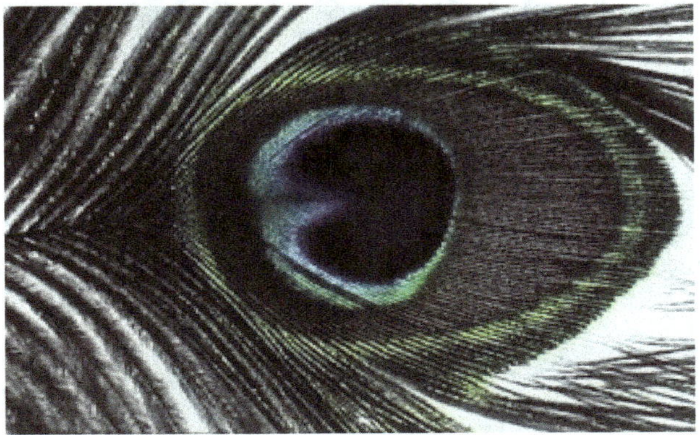

Fig. 2.10 A peacock feather

Peacock feathers are another great example of nanostructures in nature. The beautiful color of peacock feathers is due to the presence of nano sized photonic crystals on the feather [7]. Another plausible explanation for the dazzling colors is that the Peacock's feather is assumed to be coated with a film of nano dimensions and it diffracts the light giving rise to dazzling colors

2.10 Toucan Beak

The beak of the Toucan bird is very big in size when compared with its body. Inspite of this, the bird is in a position to manage to fly easily and is able to break the food items without any problem. In order understand the specialty of the beak of this bird, an electron microscopeof the beak was taken. Surprisingly, it was observed that the beak is having nanostructure andis also found to be very strong with light weight. The nanostructure of toucan beak inspires the scientists in designing automotive panels that could protect passengers in crashes and construction of strong as well as ultra-light aircraft components which will be very much useful in aerospace industries.

Fig. 2.11 Toucan

2.11 Nano Navigation Systems

Many migratory birds use magnetic structures for navigation. Homing pigeons are famous for being able to navigate extremely long distances. Homing pigeons have nanostructures in their beaks which amplify magnetic fields, making it easier for them to sense the earth's weak magnetic field.

Fig. 2.12 Home coming pigeons use nanonavigation to navigate using the Earth's magnetic field

2.12 Nanostructures in the Human Body

The human body is composed of various nanostructures such as proteins, enzymes, antibodies, bones and DNA and are essential for the normal functioning of the body. A list of nanostructures present in the human body is given in Table 2.1 [8]. In addition to this, microorganisms such as bacteria and viruses are also nanostructures that can cause illnesses in humans.

Table 2.1 List of nanostructures associated with the human body

Nanostructure	Size
Glucose	1 nm
DNA	2.2–2.6 nm
average size of protein (rubisco monomer)	3–6 nm
Hemoglobin	6.5 nm
Micelle	13 nm
Ribosomes	25 nm
enzymes and antibodies	2–200 nm

Bone Structures

The unique and contradictory properties of bone are listed below:

➢ rigid but flexible

➢ solid enough to support tissue growth with light weight

➢ porous but mechanically very strong.

➢ withstand weight without breaking.

➢ Compressive strength about twice its tensional strength.

These excellent properties might be due to their nanoscale architectural design and dimensions.

Other Nanostructures in the Human Body

Antibodies, enzymes, proteins and most organelles within cells are smaller than the micrometer-scale and a close examination of all these parts of the body clearly indicate that they are having nanostructures. Recently, lipids, self-assembled peptides, and polysaccharides were also included in the list of nanostructures present in the human body.

The birds, animals are the excellent examples of naturally available nanostructures. Taking the inspiration from these naturally available nanostructures, one may develop varietyof new materials for the benefit of mankind and as a matter of fact, these new ideas have created a new research field called **bio-inspired material science**. Further, it is possible to develop several types of new applications based on the inspiration from biological nanomaterials, represented in Table 2.2. These new materials provide new technological opportunities and potential for applications. The field of nanotechnology is still in its infancy,but if we follow nature's lead, nanotechnology and bio-mimicry have great potential to improve our society in many ways.

Table 2.2 Nanostructures inspired from biological nanomaterials

Natural systems/materials	Bio-inspired properties
Spider silk	high tensile strength fiber
wood, ligaments, and bone	high strength structural material
Eels and nervous system	Electrical conduction
deep-sea dragonfish	Transparent ceramics
Deep-sea fish and fireflies	photo emission
Butterfly wings	photonic crystals
Moth eye	anti-reflective coatings
Lotus leaf, fish scales	hydrophobic surfaces, self-cleaning
Shark skin	drag-reducing
Gecko's feet	Adhesion
Human brain	Artificial Intelligence and computing

2.13 Deoxyribonucleic Acid (DNA)

Deoxyribonucleic acid (**DNA**) is the self-replicating genetic material that exists in all living organisms, including in human beings. Apart from its role in living organism, synthetic DNA has also been studied as a novel biomaterial for various purposes ranging from biological sensors and imaging tools to analytical applications.

2.14 Applications of DNA

DNA nanotechnology provides one of the few ways to form designed complex structures with precise control over nanoscale features. Further, **DNA** along with some organic compounds are useful for biophysical studies of enzyme function and protein folding. In fact, the size of DNA is about 2.5 nm and its shape is as shown in Fig 2.13. In recent times, a series of biomolecules have been developed to cure several complex and complicated new diseases such as HIV, ebola, Zika & nipah viruses etc., which mankind has been facing during the last decade., For example, a Titanium dioxide nano particles fused with DNA gives a new Biomolecule and the drugs come out of this biomolecule may be helpful in curing some of the dreadful and incurable diseases.

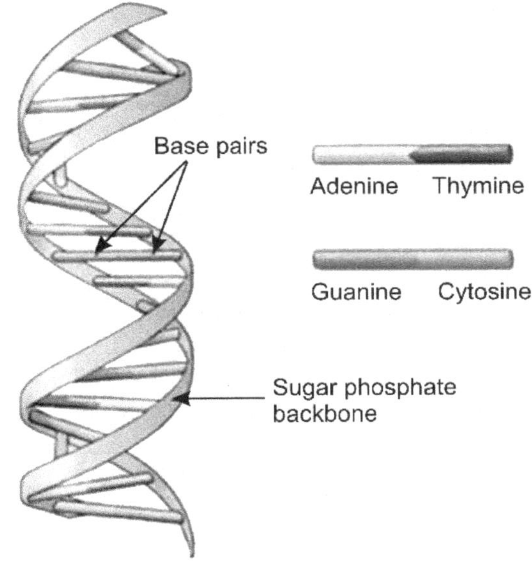

Fig. 2.13 DNA structure

Fig. 2.14 Titanium dioxide nano particles are fused to DNA

2.15 What more DNA can do?

The seed of an oak tree, acorn, uses the energy within it and reads the characteristics of a nanoscale DNA. The nanoscale DNA, gathers energy from the soil and the sun. The DNA tells the acorn to rearrange the atoms in soil, air and water to produce an oak tree. The entire process is far more complex than today's material science producing such type of structures.

References

1. Hey, Anthony J. G.; Hey, Tony; Walters, Patrick, The New Quantum Universe. Cambridge, UK: Cambridge University Press. 2003, P. 82. ISBN 0521564182.

2. M.F. Crommie, C.P. Lutz, D.M. Eigler. Confinement of electrons to quantum corrals on a metal surface. Science 1993, 262, 218-220.

3. Geometric and electronic properties of cs structures on iii-v (110) surfaces: from 1-d and 2-d insulators to 3-d metals, l.j. whitman, j.a. stroscio, r.a. dragoset, and r.j. celotta, phys. rev. lett., 1991, 66, 1338.

4. Nanomaterials for the Life Sciences Vol.7: Biomimetic and Bioinspired Nanomaterials.Edited by Challa S. S. R. Kumar, WILEY-VCH Verlag GmbH & Co. KGaA,Weinheim, 2010, ISBN: 978-3-527-32167-4

5. Yoshioka, S.; Kinoshita, S. Wavelength-selective and anisotropic light-diffusing scaleon the wing of the Morpho butterfly. *Proceedings: Biol. Sci.* 2004, 271(1539): 581- 587. DOI:10.1098/rspb.2003.2618.

6. Balasubramanian Karthick, Ramesh Maheshwari, Lotus-inspired nanotechnology applications, Resonance, 2008, 13(12):1141-1145, DOI:10.1007/s12045-008-0113-y

7. Jiyu Sun, Bharat Bhushan and Jin Tonga, Structural coloration in nature, RSC Adv., 2013, 3, 14862–14889, DOI: 10.1039/c3ra41096j

8. Jaison Jeevanandam, Ahmed Barhoum, Yen S. Chan1, Alain Dufresne and Michael K.Danquah, Review on nanoparticles and nanostructured materials: history, sources,toxicity and regulations, Beilstein J. Nanotechnol. 2018, 9, 1050–1074. doi:10.3762/bjnano.9.98.

Chapter 3

Classification and Surface to Volume Ratio in Nanomaterials

3.1 Quantum Confinement Effect

The quantum confinement is the restriction of the motion of randomly moving electrons present in a material to a small region, whose dimensions are of the order of wavelength of an electron, called as de Broglie wavelength. When the electrons are confined to such a small region, their energy and momentum are not wasted. Therefore, the scenario is now that we have electrons confined in the conduction band and holes in the valence band. The motion of electrons in the confined to a very small space can be modeled by the motion of a particle in a potential well with infinite walls. The phenomenon of **quantum confinement** normally observed in nano-materials [1].

Broadly speaking, the quantum confinement is the restriction of the motion of electrons present in a material to a small region. As the electrons are totally confined to a small region, it may result in controlling their energy which may lead to the creation of discrete energy levels rather than to quasi continuum of energy bands. It is an effect where the dimensions of a typical material are comparable with those of de Broglie wavelength of the electrons involved. At such low dimensions, the electrons present in the material behave more or less similar to the electrons present in atoms. In other words, electrons occupy discrete energy levels rather than a quasi-continuum of energy in a band. When the material is in sufficiently small in size typically 10 nm or less ($\leq$ 10nm), energy levels of electrons change considerably. This may result in the confinement of "electrons" and "holes" into a small region and the phenomenon is called quantum confinement. This phenomenon in turn influences the electrical, optical, and magnetic behavior of materials.

Quantum confinement is the spatial confinement of electrons of nanomaterial in one or more dimensions within a material. The overall behavior of bulk crystalline materials changes when the dimensions are reduced to the nanoscale. Based on the confinement direction, a quantum confined structures maybe classified into three categories as quantum dots, quantumwire and quantum well.

3.2   Classification of Nanomaterials

Nanomaterials are of great interest because of their different physical as well as chemical properties when compared with their bulk counterparts. In order to evaluate the properties of nanomaterials, appropriate classification is needed. The most commonly accepted ways for classifying nanomaterials is based on their dimensions. Therefore, based on their spatial dimensions and also on the basis of quantum confinement, nanomaterials may be classified as [2],

➤ Zero dimensional (0-D) nanomaterials

➤ One dimensional (1-D) nanomaterials

➤ Two dimensional (2-D) nanomaterials

➤ Three dimensional (3-D) nanomaterials

and all the details of the classification are given in Table 3.1.

Table 3.1 Classification of nanomaterials

Dimension of the nanomaterial	Examples	Number of free dimension	Quantum confinement
Zero Dimensional material	Quantum dot(0-D)	0	3
One Dimensional material	Quantum wire (1-D)	1	2
Two Dimensional material	Quantum wells (2-D)	2	1
Three Dimensional material	Bulk (3-D)	3	0

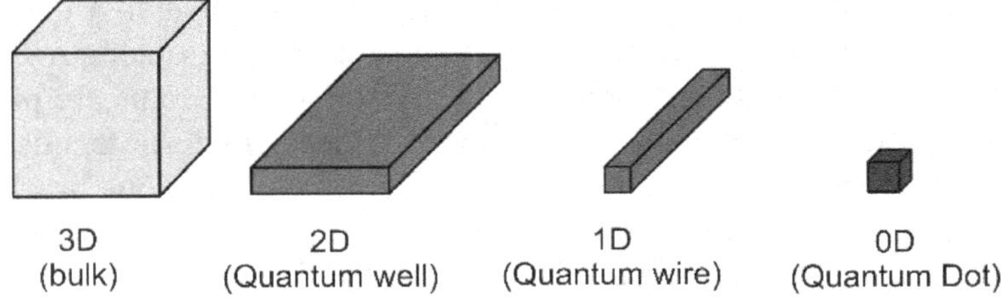

Fig. 3.1 Classification of nanomaterials

3.3 Zero Dimensional (0D) Nanomaterials

Materials which have all the three dimensions in the nanoscale range (no dimensions or 0- D, are larger than 100 nm) and if the electron is confined in all three dimensions are called as zero dimensional (0-D) nanomaterials. In these materials, no electron delocalization (freedom to move) takes place. The most common representation of zero dimensional (0-D) nanomaterials is quantum dots. These nanoparticles can be:

➢ Amorphous or crystalline

➢ Single crystalline or polycrystalline states

➢ Metallic, ceramic, or a polymeric

➢ Composed of single or multichemical elements

➢ Exhibit various shapes and forms

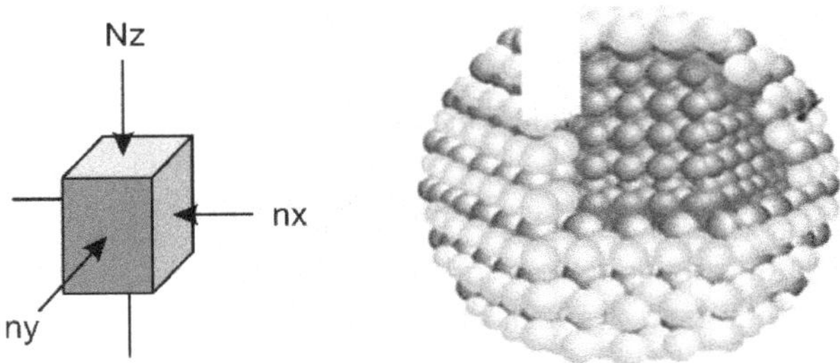

Fig. 3.2 Zero dimensional materials

3.4 Quantum Dots (QDs)

Among the available zero dimension materials, the Quantum dots (QDs) are the most studied materials. Quantum dots are tiny semiconductor particles with their size ranging from 1-5 nm. These materials are having exceptional optical and electronic properties that differ from their bulk counterparts probably due to quantum confinement [3]. These materials obey the principles of quantum mechanics. The energy band gap increases with a decrease in size of the quantum dots. Because of smaller band gap, even the UV light, is enough to excite an electron in quantum dots to higher energy levels. In a quantum dots, relatively few atoms are present, and are confined to a much smaller space. This leads to discrete, quantized energy levels more like those of an atom than the continuous bands of a bulk semiconductor. For these reasons, quantum dots are sometimes referred to as "artificial atoms." The properties of a quantum dots are not only determined by their size but also by their shape, composition, and structure, for instance if it's solid or hollow. These wonderful materials are having enormous applications in catalysis, electronics, photonics, information storage, imaging, medicine, sensing, optoelectronics and manufacturing of inks and paints for anti- counterfeit tagging.

Other examples of zero dimensional materials category includes Bucky ball (C60)

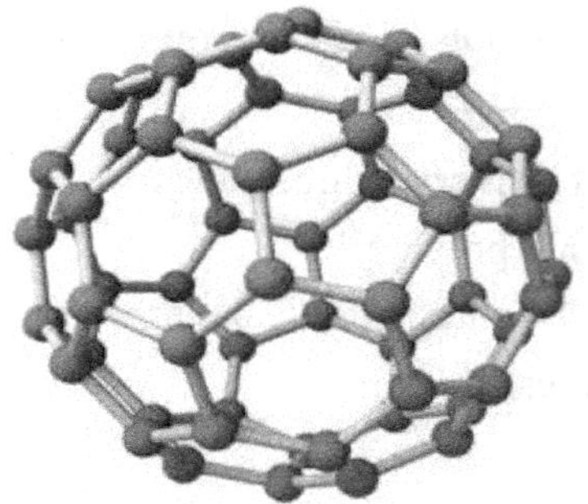

Fig. 3.3 Bucky ball (C60)

3.5   One Dimensional (1-D) Nanomaterials

For 1-D nanomaterials, electron confinement occurs along two dimensions [4], whereas delocalization takes place along the long axis of this material.

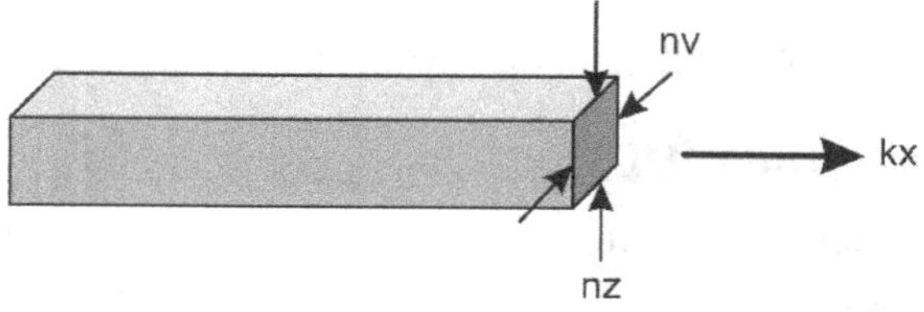

Fig. 3.4 One dimensional Nanomaterial

Quantum wires and Carbon nanotubes are the best example for 1-D nanomaterials. Other examples of this category include nanotubes, nanorods, and nanowires. Similar to zero- dimensional nanomaterials, one-dimensional nanomaterial can also be:

➢ Amorphous or crystalline

➢ Single crystalline or polycrystalline states

➢ Metallic, ceramic, or a polymeric

A single wall carbon nanotube (SWNT) is a one-atom thick graphite sheet rolled up into a cylinder with diameter of the order of a nanometer. Carbon nanotubes are the strongest and stiffest materials in terms of their tensile strength and elastic modulus. The strength and flexibility of carbon nanotubes makes them potential materials for several applications.

Ex: Carbon nanotubes (CNTs).

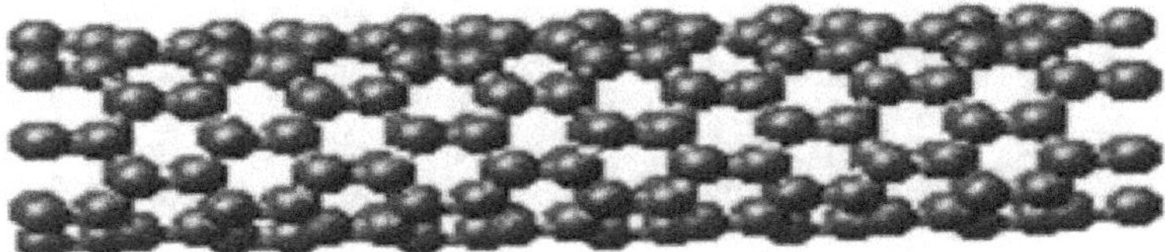

Fig. 3.5 Carbon Nanotube

3.6   Two Dimensional (2-D) Nanomaterials

Materials in which any two of three dimensions are not confined to the nanoscale range are called two dimensional (2-D) nanomaterials. These materials exhibit plate-like structures and some of the examples of this category include nanofilms, nanolayers, nanocoatings, graphene, 2-D quantum well etc,. In these materials, as the conduction electrons are confined across the thickness they are delocalized in the plane of the sheet. As more number of the dimension is confined, more discrete energy levels can be found, in other words, electron movement is strongly confined in a given dimension.

This class of materials can be:

➢ Amorphous or crystalline

➢ Single crystalline or polycrystalline states

➢ Made up of various chemical compositions

➢ Metallic, ceramic, or a polymeric

➢ Deposited on the substrate

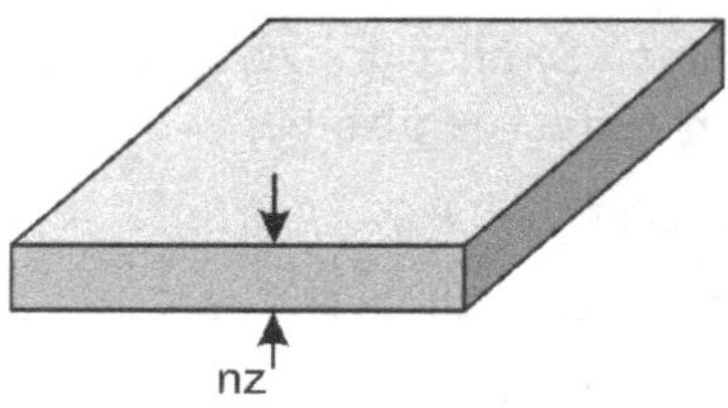

Fig. 3.6 Two dimensional Material

Fig. 3.7 Graphene layer [5]

Graphene is an allotrope of carbon in the form of a single layer of atoms in a two- dimensional hexagonal lattice. Graphene has a special set of properties like it is about 100 times stronger than the strongest steel. However, its density is dramatically lower than any of the steels. It conducts heat and electricity very efficiently and is nearly transparent.

3.7   Three Dimensional (3-D) Nanomaterials

Materials in which all the three dimensions are not confined to the nanoscale range are called as three dimensional (3-D) nanomaterials. Three dimensional nanomaterials also known as bulk nanomaterials. Three dimensional (3-D) nanomaterials are aggregations of any of the above (0-D, 1-D or 2-D) dimensional nanomaterials with greater than 100nm in all the three directions. Three dimensional (3-D) nanomaterials can contain dispersions of nanoparticles, bundles of nanowires and nanotubes as well as multiple nanolayers. This class of materials can be:

➢ Amorphous or crystalline

➢ Made up of various chemical compositions

➢ Metallic, ceramic, or a polymeric

➢ Composed of multiple nanolayers

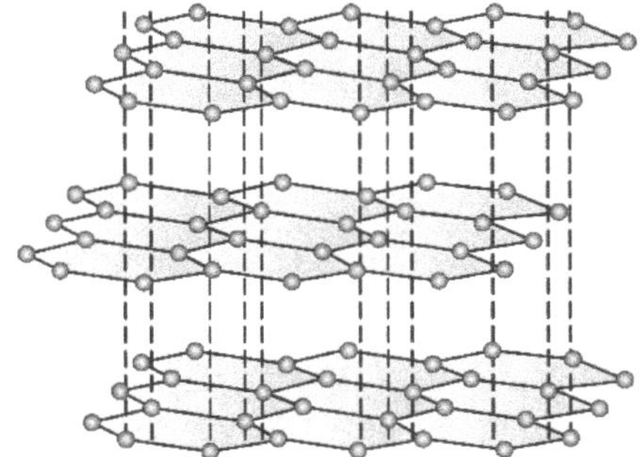

Fig. 3.8 Graphite

Fig. 3.9 Multiwall Carbon Nanotube

Conclusion: Finally, one may conclude that for 0-D nanomaterials the electrons are fully confined. On the other hand, for 3-D nanomaterials the electrons are fully delocalized. In case of one and two-dimensional nanomaterials, electron delocalization and confinement coexist. This confinement significantly influences the conduction band of the material. The impact of quantum confinement on the resultant energy level diagram of the material and can be analyzed using quantum mechanics. For example, consider an electron present inside an infinite height potential box. The electron cannot escape from this quantum well and is confined by the dimensions of the nanostructure. In the case of 0-D, 1-D and 2-D nanomaterials the effect of confinement on the energy states can be written respectively.

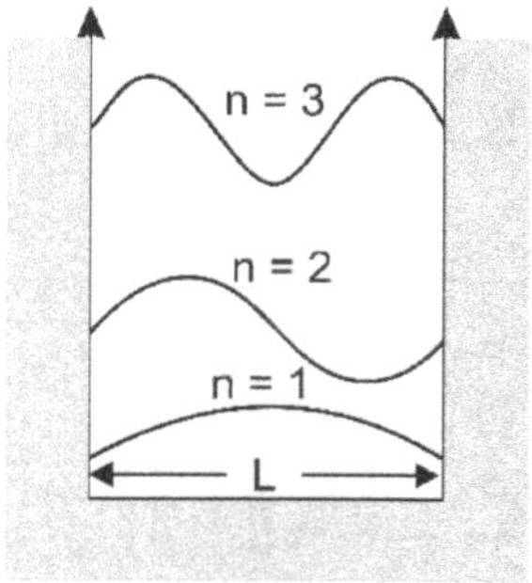

$$(0\text{-}D) \qquad E_n = \left[\frac{\pi^2 h^2}{2mL^2}\right]\left(n_x^2 n_y^2 + n_z^2\right)$$

$$(1\text{-}D) \qquad E_n = \left[\frac{\pi^2 h^2}{2mL^2}\right]\left(n_x^2 + n_y^2\right)$$

$$(2\text{-}D) \qquad E_n = \left[\frac{\pi^2 h^2}{2mL^2}\right]\left(n_x^2\right)$$

Fig. 3.10 Electron in a potential well and the energies are shown on the right

Where $\hbar = h/2\pi$,

h is Planck's constant

L is the width or length (confinement) of the potential well,

m is the mass of electron,

n_x, n_y, and n_z are the principal quantum numbers with respect to x, y, and z dimensions.

As the size of the nanostructure decreases to a smaller and smaller value of L, the gap between the energy levels increases resulting in discreet energy spectrum. Therefore, energy band structure in the nanomaterials is quite different from their bulk counterpart and this in turn alters most of the physical and chemical characteristics of these materials.

3.8 Size Effects

Nanomaterials are of great interest because the properties of *materials* change drastically as their size approaches the nanoscale. The principle factor cause the properties of nanomaterials differ significantly from the other materials is increase in surface to volume ratio(S/V) which is explained below.

3.9   Surface Area to Volume Ratio

One of the most fundamental differences between nanomaterials and bulk materials is that nano scale materials have very large value of surface to volume ratio. The larger surface area of nanomaterials plays a major role in dictating these material's important properties [6]. The surface area to volume ratio and its effect on the properties of the nanomaterials is the key feature of nanoscience and nanotechnology. To understand this concept, let us see some examples.

Ex 1:

Consider a spherical material of radius 'r'

The surface area of the sphere(S) $\quad = 4\pi r^2$

The volume of the sphere (V) $\quad = 4/3\pi r^3$

Surface area to Volume ratio(S/V) $\quad = \dfrac{4\pi r^2}{4/3\,\pi r^3}$

$\therefore$ Surface area to Volume ratio $\quad = \dfrac{S}{V} = \dfrac{3}{r}$

It means that the surface area to volume ratio increases with decreasing diameter of the sphere and vice versa.

Ex 2:

Next, consider a material in cylindrical shape of radius 'r' and height is 'H'In this case, Surface area (S) = $2\pi rH$

Volume (V) = $\pi r^2 H$

Surface area to Volume ratio(S/V) $\quad = \dfrac{2\pi rH}{\pi r^2 H}$

$\therefore$ Surface area to Volume ratio $\quad = \dfrac{S}{V} = \dfrac{2}{r}$

The trend is similar to the sphere case, as the size reduces, the surface area to volume ratio increases.

Ex 3:

Let us now consider a material with cubic shape, if the side of the cube is 'L'Surface area (S) = $6L^2$

Volume (V) = L^3

Surface area to Volume ratio(S/V) $\quad = \dfrac{6L^2}{L^3}$

$\therefore$ Surface area to Volume ratio $\quad = \dfrac{S}{V} = \dfrac{6}{L}$

In this case also, the surface area to volume ratio is inversely proportional to the size of the material. If the material is a cube and as it is divided into small cubes as shown below figure, then also the surface area to volume ratio increases. It also means that when a given volume of material is made up of smaller particles, the surface area of the material increases. Hence, the nanomaterials possess large value of surface area to volume ratio as compared to the bulk material.

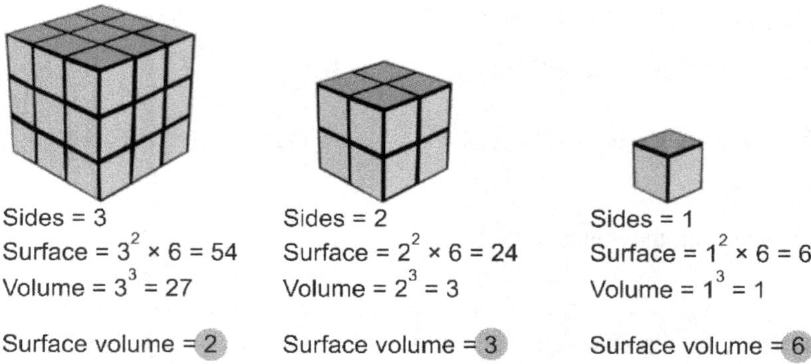

Fig. 3.11 Explanation of Surface to Volume ratio phenomenon [7]

3.10 Effect of Surface Area to Volume Ratio

Materials with smaller dimensions (nano dimension) have a much greater surface area to volume ratio (S/V) when compared with their bulk materials counter parts. This may result   in making the particles on the surface very active. In fact, as the particle size reduces, a large no. of atoms present will be coming on to the surface. Therefore, the nanoparticles become very active. That means the materials which are chemically inactive (inert) in their bulk form are chemically active in their nanoparticle form. Therefore, the properties of nanomaterials are drastically different from those of bulk materials because of their small dimensions, nanomaterials have a vast surface area to volume ratio (S/V). Moreover, this results the large fraction of atoms of the material be on the surface, resulting in more surface dependent material properties.

In bulk materials, only a small percentage of crystallites are there near the surface (grain boundary). However, in nano materials, small crystallite size ensures that many of them perhaps half or even more in some cases, are near surface.

For example,

In the case of a 30 nm iron nanomaterial, 5% atoms are present on the surface, while in the case of a 10 nm nanomaterial of the same compound, 20% of the atoms present on the surface and in the case of a 3 nm nano particle 50% are present on the surface. Therefore, surface to volume ratio is having profound effect on various properties of nanomaterials.

References

1. A. I. Onyia, H. I. Ikeri, A. N. Nwobodo, Theoretical study of the quantum confinement effects on quantum dots using particle in a box model, Journal of Ovonic Research Vol.14, No. 1, 2018, pp. 49 – 54

2. Nanoengineering Global Approaches to Health and Safety Issues, edited by Patricia I. Dolez, Elsevier, 2015, pp 6-7

3. D. Sumanth Kumar, B. Jai Kumar, H. M. Mahesh, Synthesis of Inorganic Nanomaterials, Advances and Key Technologies, Micro and Nano Technologies, 2018, Pages 59-88.

4. Edvinsson T., Optical Quantum Confinement and Photocatalytic properties in two, one and zero-dimensional nanostructures. R. Soc. open sci., 2018, 5:180387.
https://innovationtoronto.com/2016/04/graphene-layer

5. Manzoor Ahmad Gatoo, Sufia Naseem, Mir Yasir Arfat, Ayaz Mahmood Dar, Khusro Qasim, and Swaleha Zubair, Physicochemical Properties of Nanomaterials: Implication in Associated Toxic Manifestations, BioMed Research International, 2014; Vol. 2014, 498420.
https://sustainable-nano.com/2014/09/23

Chapter 4

Synthesis of Nanomaterials

4.0 Introduction

Nano science & Nanotechnology are playing a pivotal role in the development of science and technology. In recent times, nanomaterials have huge demand due to their novel properties and a large number of applications in various fields. Many synthesis techniques are available for the preparation of nanomaterials. The preparation technique decides the quality and the characteristics of the resultant material. Therefore, the selection of a suitable synthesis technique is important in obtaining good quality nano particles.

4.1 Synthesis Technique

Although there are several routes for the preparation of nanomaterials, the routes of synthesis are broadly classified as [1-2],

(i) Top-Down Technique

(ii) Bottom-up Technique

4.2 Top-Down Technique

In the case of the Top-down method, the bulk material is brake down into minute particles until the desired nanoscale material is obtained. This approach may involve mechanical methods such as cutting, carving and moulding. A simple way to elucidate a top-down approach is to think of carving a statue out of a large block of marble. Nanolithography and mechanical alloying methods also belong to this category. Top-down method has various advantages like it has a great potential of developing a variety of complex nanostructures with a large scale production, cost-effective method and chemical purification also not required. The method is explained in a simple and lucid manner pictorially in Fig.4.1.

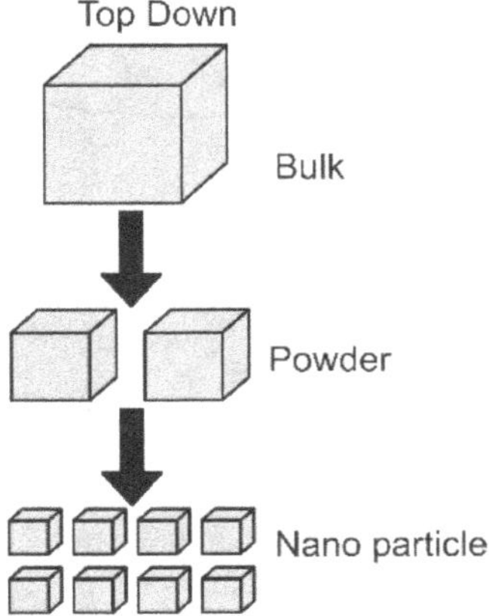

Fig. 4.1 Top-Down approach

4.3   Solid State Reaction Method

Solid state reaction (SSR) method is the oldest and one of the most commonly used methods for preparing multi component solids. In this method, the starting materials are taken in the form of oxides or carbonates. After weighing the starting materials in stoichiometric ratio, they are mixed thoroughly and uniformly. Since, solids do not react with each other at room temperature, it is necessary to heat them at very high temperatures of as high asup to 1500 °C. At these high temperatures, the reaction takes place at an appreciable rate also. The precursor mixture is sintered for at a specific temperature rate so as to facilitate solid-state chemical reaction among the starting chemicals and the compound is formed. The final temperature and duration of sintering may vary depending on the nature and properties of the sample under preparation. Synthesis of ceramic powders by the conventional solid-state reaction method and repeated heating with intermediate mechanical or hand-grinding process is almost always essential to produce a single-phase powder to attain chemical homogeneity. Thus, both the thermodynamic and kinetic factors are important in solid-state reaction (SSR) method.

The advantages of SSR method is the easy availability of the precursors, simplicity of the process, low cost preparation methods, low cost laboratory equipment and easy to upscale the industrial production. The major advantage of SSR method is that the final product is in the solid state form, the structural purity of the compound depends on the final sintering temperatures. Moreover, this method is environmental friendly and no toxic or unwanted waste is produced from this method [3]. A flow chart diagram of various steps involved in this method is given in Fig.4.2.

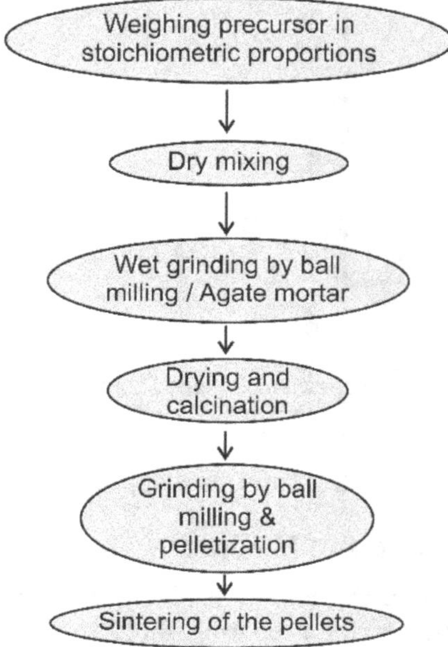

Fig. 4.2 Flow chart diagram of Solid State Reaction method

The solid state method is also having some drawbacks. They include,

➤ Poor compositional control

➤ Chemical in homogeneity

➤ During grinding there is probability of contamination with impurities

➤ The method requires very high temperature ($>1000^{0}C$)

➤ Large particle size

4.4   Hydrothermal Synthesis

Hydrothermal method is one of the most popular techniques for the synthesis of nanoparticles. This method is a solution-based reaction method [4]. In the hydrothermal method, to control the size and shape of the resultant nanoparticles, a wide range of temperatures i.e. from normal to high temperatures and different ranges of pressures i.e., from low-to high-pressure conditions can be used based on the vapour pressure of the main chemical composition in the reaction.

In this method, the precursor materials are dissolved in a solvent (aqueous solution) and transferred into a pressurized vessel called an autoclave, the vessel sealed tightly and subjected to thermal treatments. The temperature in the autoclave can be raised above the boiling point of water. In such conditions, high pressure is built up inside the autoclave. Under these high pressure conditions, the boiling point of aqueous solution used in the auto clave goes up considerably. For example, the B.P of water (B.P of water at NTP is 100^0 C) goes beyond 300^0 C, thereby increasing the temperature of the solution used in the auto clave. This in turn promotes the chemical reactions between the precursors and the autoclave solution thereby producing nano powders. As the pressure inside the reaction vessel increases, significant changes on the physical properties of the solvent are observed over a wide temperature.

Finally, when the temperature and pressure attain "critical values" (T_C = 374°C, P_C = 218 atm), a new phase called a supercritical state of water has formed. In this state, water (solvent) exhibits both characteristics i.e., liquid and gas phases making it a very active medium for varied chemical reactions. Under such extreme conditions, chemical reactions can occur in both sub-critical and super-critical regime, thereby resulting into highly crystalline single-phase products which do not require post-synthesis annealing. Hydrothermal method in supercritical state of water has advantages for the preparation of metal oxide compounds because the rate of reaction is improved greater than 1000 times that of the normal hydrothermal synthesis conditions. In order to increase the crystallization kinetics, it is necessary to introduce the microwave, electric or ultrasonic fields in the hydrothermal systems and are referred to as microwave- hydrothermal, electrochemical hydrothermal and ultrasonic hydrothermal, respectively which contribute to make this method expensive and time consuming.

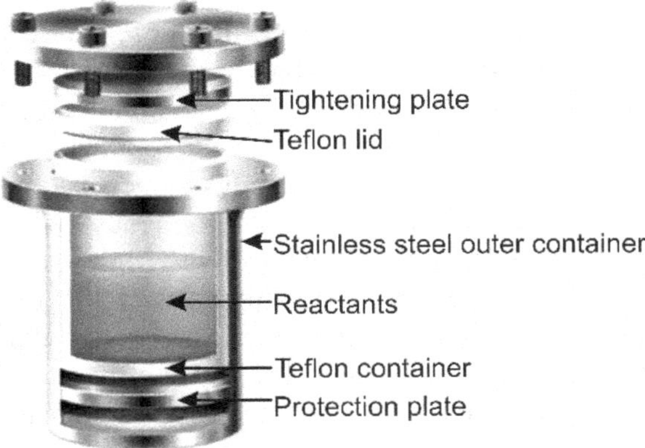

Fig. 4.3 Schematic diagram of the autoclave with Teflon liner [5]

Advantages of hydrothermal process:

➢ The hydrothermal process is highly efficient "green-chemistry route.

➢ The major advantage of this approach is that most of the inorganic materials can be made soluble in water at elevated temperatures and pressures.

➢ The hydrothermal environment offers a lower viscosity, higher diffusivity and higher dissolving power for efficient synthesis of nanomaterials

➢ It is possible to have higher reaction rate between the precursors.

➢ This method facilitates low temperature preparation of nanomaterials.

➢ Phase purity and homogeneity is possible in this method.

Limitations

➢ The method is expensive because of the manufacturing cost of autoclave itself is costly.

➢ A lot of safety precautions are to be followed.

➢ The entire reaction process cannot be observed in this method.

➢ This method allows only for a limited control of the nucleation and growth processesusually yielding particles with irregular shapes.

➢ The reaction time, temperature and dielectric constant of the reaction medium may exert a play a role on the particle size and the crystallinity of the nanomaterials.

4.5   Solvothermal Synthesis

A chemical reaction that is carried out in the presence of a non-aqueous solvent is commonly referred as solvothermal reaction. Solvothermal synthesis is widely carried out methodology for the synthesis of various types of nano materials, including metals, semiconductors, ceramics, and polymers etc,. The route is quite similar to hydrothermal synthesis, except the precursor materials used are all non-aqueous solutions. In this synthesis, the solvent is maintained under low to high pressure (from 1 to 1,000 atm) and a wide range of temperatures (from 100 to 1000 °C) that promotes the chemical reactions between the precursors during synthesis. Additionally, this technique ensures the precise control over the size, shape distribution and crystallinity of the metal oxide nano particles or nano structures synthesized. Moreover, the characteristics features of the synthesized material can be varied by adjusting the experimental parameters such as reaction temperature, duration of the reaction, nature of the solvent, surfactant and precursor used.

4.6 Nanolithography

Lithography is a sophisticated process, through which the components essential for semiconductor devices such as integrated circuits (IC's), micro electromechanical systems (MEMS), optoelectronic components and displays are fabricated. In fact, was the main technique used in microelectronics industry for the last several years. These developments in the semiconductor and IC industry has led to a new paradigm of the information technology via computers and the internet. In fact, for decade's lithography techniques are in use for the manufacture of integrated circuits (ICs) and microchips

After the discovery of nanotechnology, people started using lithography technique for variety of nanotechnological applications and as such the technique is referred as nanolithographic technique. The micro and nanolithography technologies were used to create patterns with a size ranging from a few nanometers up to tens of millimeters [6]. Thus, Nanolithography deals with the study of the preparation of nano structures and is one of the several techniques available for the preparation of nanosize materials. Finally, Nanolithography is the science of etching, writing or printing to modify a material surface with structures under 100nm.

Steps followed in the Lithographic process

Following steps are followed in the basic lithography process.

1. Identification of a suitable substrate material is the first task.
2. Using a suitable technique, main material is to be coated on the substrate.
3. The main material gives the required properties.
4. On the main material, a thin layer of photosensitive material is to be coated.
5. Now both the main material and photosensitive materials are to be masked with a suitable masking material.
6. Now the entire arrangement is to be exposed with UV light or X-rays.
7. Because of the exposure to UV light, the required pattern in the photosensitive material shown in the figure is developed.
8. Later etch the photo sensitive material so that the required pattern of the main material is emerged.
9. Finally remove the unwanted photosensitive material.

Limitations

Although the concept and process of nanolithography is simple, the process sometimes may create problems and are given below:

➢ The implementation is itself a very complex process.

➢ Nano structures less than 100nm are difficult to produce due to diffraction effects.

➢ Masks need to be perfectly aligned with the pattern on the silicon vapor.

➢ The silicon vapor is to be a perfect single crystal, almost defect free.

➢ Nanolithographic tools are very expensive.

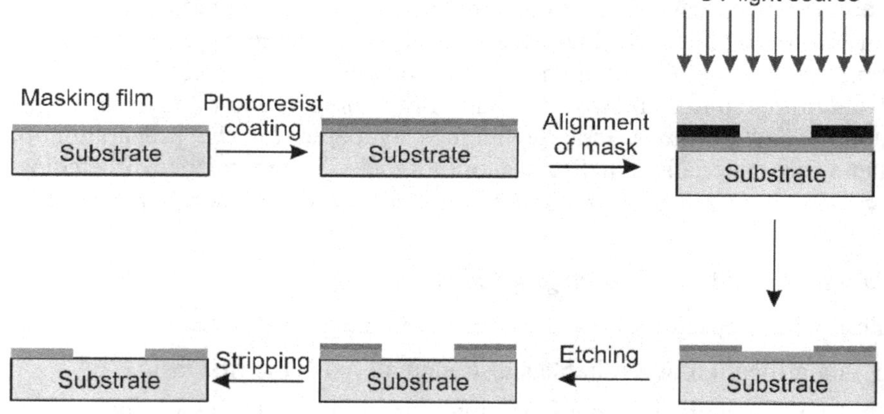

Fig. 4.4 Schematic representation of a photolithography process [7]

4.7   Bottom-Up Techniques

The formation of nanomaterials via atom-by-atom, molecules-by-molecules or cluster-by-cluster approach is called a bottom- up method. In fact; the basic principle of the bottom-up methods is an opposite approach to the top-down methods. In the bottom-up method, the nanomaterials are obtained starting from the atomic or molecular precursors and gradually assemble them until the desired nanostructure is formed. The method resembles the construction of a big building by brick by brick. The bottom-up methods play a significant role in the field of nanotechnology for designing, developing, growth of nanostructures and synthesis of nanomaterials. These bottom-up techniques are able to produce various nanostructures like nanoparicles, nanonanowires, nanotubes, nanorods and nanofilms etc. with different size, shape, chemical composition, morphologies, and structures. Solution based approaches like the sol-gel method, molecular self-assembly, micro-emulsion process, co-precipitation, atomic layer deposition (ALD) and chemical vapour deposition (CVD) belong to the category of bottom - up techniques for synthesis of nanomaterials. The diagrammatical representation of the process is shown in Fig. 4.5. A comparison between the two methods is also explained in Table 4.1.

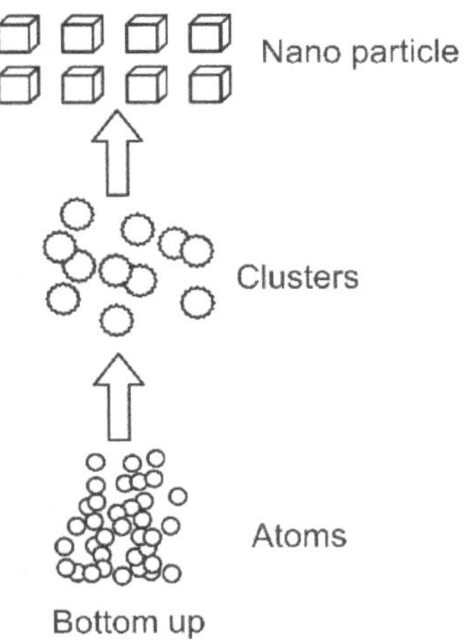

Fig. 4.5 Bottom - Up Method

Table 4.1 Comparison between Top-Down and Bottom-Up techniques

Top-Down technique	Bottom- Up technique
Advantages	
➢ Large scale material production ➢ Thin film deposition over the large surface area is possible ➢ Purification of raw materials (chemicals) is not necessary	➢ Ultra-fine and homogeneous nanoparticles, nanowires, nanotubes, thin films can be prepared ➢ Narrow and uniform particle size distribution ➢ Simple and less expensive method ➢ Quick processing time ➢ Minimum imperfections

Table 4.1 *contd...*

Disadvantages	
➤ Broad particle size distribution ➤ Varied particle size and shapes ➤ stress, defects and contaminations(impurities) during the preparation of the nanostructures ➤ Expensive technique	➤ Large scale production is difficult ➤ Chemical purification is required

4.8   Sol-Gel Method

In Materials Science, the sol–gel process is a method for producing solid materials starting from atoms or molecules. The method is therefore called as the bottom- Up process. The method is used for the preparation of oxide based materials. The process involves conversion of monomers into a colloidal solution (sol) that acts as the precursor for an integrated network (or gel) of either discrete particles or network of polymers. In fact, the sol–gel method is a wet-chemical technique used for the fabrication of nanomaterials in the form of nanoparticles, nanofibers, nanofilms, aerogels and nanoporus materials.

Sol: "Sol" is the uniform dispersion of the solid particles in a homogeneous liquid medium (solvent) in which the particles are suspended only due to the Brownian motions.

Gel: The gel is a phase in which both solid and liquid components are dispersed in each other.   It exhibits a solid network that has liquid components.

Synthesis procedure

Various steps involved in the synthesis of the nanoparticles by sol-gel method are explained below [8]

Sol-formation (Hydrolysis)

The precursors have taken in stoichiometric ratios and are completely dissolved in appropriate quantity of de-ionized water (or alcohol). In the process, the hydrolysis reaction takes place, wherein the -OR group of the precursor chemical is replaced with an -OH group. Normally, in any hydrolysis reaction a base (NaOH or NH3) or an acid (HF or CH3COOH) is used as a catalyst. In contrast, the hydrolysis reaction of sol-gel route, takes place without a catalyst and reaction is more rapid.

Typically, the starting materials are metal alkoxides or metal nitrates used to form solvated metal precursors (sol). The precursors are hydrolyzed with water to produce hydroxide as shown in the following Eq. (4.1).

$$M(OR) + H_2O \leftrightarrow M(OH) + ROH \qquad \qquad ....(4.1)$$

Gel formation (Condensation)

The next step in the sol-gel process is the stirring. The stirring process is continued at least for one hour. The solution is evaporated due to intensive stirring at a temperature of 100°C and heating process is continued until the sol turned into a viscous gel. The hydroxide molecules later form oxide polymer network (the gel) through condensation reactions (condensation reaction is shown in Eq. (4.2). At this moment, gelation occurs. A gel consists of a three-dimensional network with interconnected pores. The rate of hydrolysis and condensation strongly influence the characteristics of the products. A slower, careful and more controlled hydrolysis process will give uniform and small size particles.

$$M - OR + M\text{-}OH \rightarrow M{-}O{-}M + ROH \qquad \qquad ....(4.2)$$

Drying

Now stirring is stopped but heating is continued until the entire gel is burnt into powder.

Dehydration

During this process, the surface bound M-OH groups are removed. This is normally achieved by undertaking the calcination process normally conducted between 600-800^{0}C. In fact the calcination process may be useful in reducing lattice defects and strains. In fact, the chemical reaction between various components takes place during this process. Subsequently, the resulting powder is crushed in an agate mortar for removing the pores from the sample and to obtain fine powder of nano particles.

Final Sintering

Finally, the powders after pelletizing into a particular shape are sintered in a furnace between 800- 1000^{0}C. This process also helps in densification of the samples. The typical steps that are involved in sol-gel processing are shown in the schematic diagram below.

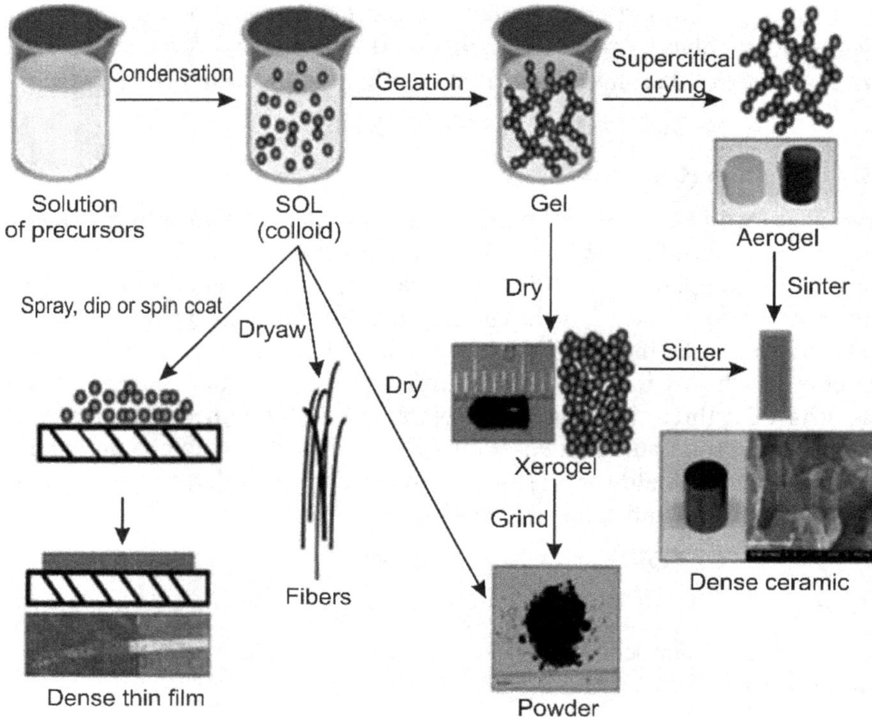

Fig. 4.6 Schematic diagram of obtaining nanomaterials by sol-gel process [9]

Advantages of Sol-Gel technique

➢ The sol-gel process is cost effective, simple and easy process.

➢ Precise control over phase and chemical constituents is possible.

➢ The final products are relatively homogeneous and also highly pure.

➢ Fine particle size and uniform particle size distribution is possible.

➢ Desired material can be synthesized at lower temperatures.

➢ Samples prepared by this method might have various shapes and dimensions that include films, fibers and powders.

4.9   Chemical Vapour Deposition (CVD)

Chemical vapor deposition (CVD) method is used to prepare good quality solid materials. This approach is commonly used in the semiconductor industry to develop nanofilms, for a number of electronic and optoelectronic applications [10].

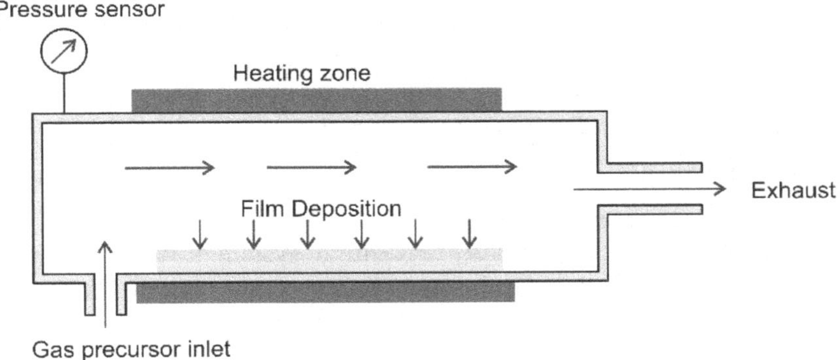

Fig. 4.7 Schematic diagram of a chemical vapor deposition (CVD) system

CVD setup consists of five important components. They are,

1. Precursors delivery system
2. Reaction chamber
3. Energy source
4. Vacuum system
5. Exhaust gas handling system

Precursors Delivery System

In CVD technique, the precursor is delivered into the reaction chamber, in the form of a solid or a liquid or a gas. In order to deliver the precursor in a controlled manner, CVD requires the precursor's delivery systems. A typical precursor delivery system generally consists of a control meter called flow meter, which controls the flow rate of the precursor. In this case the precursor (material) is heated to boiling point in order to generate precursor vapor that could be transported by carrier gas into the reaction chamber.

Reaction Chamber

In the CVD set up the reaction chamber is the heart of the system. Some main parts of the CVD chamber are given below

➢ Quartz or Alumina tube.

➢ Couplings to maintain the required pressure in the chamber

➢ Inlet and Outlet for the gas flow into the reaction chamber

Energy Source

There are various energy sources available to supply heat in the CVD reaction Chamber which includes radiant heating, resistance heating, laser heating, rapid thermal annealing, electric induction heating, magnetic induction heating, etc. For materials growth, temperature uniformity within the chamber is one of the key parameters to be precisely controlled for CVDprocess depending upon requirement.

Vacuum System

The CVD process is usually carried out in vacuum conditions to eliminate the atmospheric contaminations. A vacuum system is usually used in CVD to provide continuous and uniform pressure throughout the process. This helps in growing homogeneous and better quality materials.

Exhaust Gas Handling Systems

The gas handling system is used to remove the byproduct gases that are created during the reaction process from the reaction chamber and to make clean, non-toxic and hazardous byproducts before injecting into the open atmosphere.

Working principle of CVD

In CVD process, the vaporized precursors are introduced into a CVD chamber and made to adsorb by a substance held at higher temperature, called substrate. The adsorbed molecules react on the substrate surface and produce the desired material. Thus the three main steps involved in the CVD process are -Transportation of the reactants onto substrate, the chemical reactions on the substrate and Removal of byproducts formed by the gas-phase reactions. By varying the experimental conditions substrate material, substrate temperature, composition of the reaction gas mixture, total pressure gas flows, etc. materials with different structures can be grown in this method.

Advantages:

➤ Uniform distribution of the sample and high growth rate

➤ Selective and large area deposition is possible

➤ High purity products can be obtained.

➤ CVD films are harder than similar materials produced using conventional ceramic fabrication processes.

➤ Excellent reproducibility

➤ The production films by CVD method is economically viable.

Disadvantages:

➤ Highly corrosive, inflammable, explosive and toxic precursor Chemicals may create problems and as such all sorts of precautions are to be taken.

➢ As the by-products of CVD process are toxic and their neutralization is highly expensive

➢ A lot of safety and contamination related aspects have to be addressed

➢ High temperatures (often greater than 600 °C) may not be suitable for all the purposes. High purity chemicals used in this process with high cost are required to obtain good quality films.

4.10 Physical Vapor Deposition (PVD)

Physical vapor deposition (PVD) is the vacuum deposition methods which can be used to produce thin films and coatings [11]. In fact, this technique is an environmental friendly vacuum deposition technique consisting of three fundamental steps:

1) Vaporization of the material from a solid source

2) Transportation of the vaporized material

3) Condensation of the vapor on the substrate.

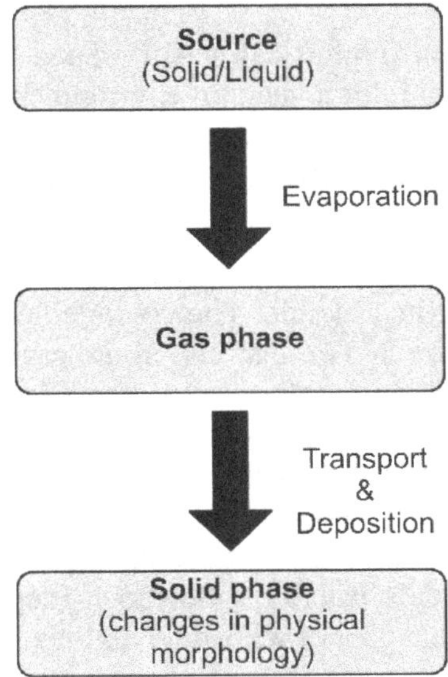

Fig. 4.8 Steps in PVD [12]

Physical vapor deposition (PVD) is a thin-film coating process which produces coatings of pure metals, metallic alloys and ceramics. Physical vapour deposition, as its name implies, involves physically depositing atoms, ions or molecules of a coating species on to a substrate. Different PVD technologies utilize the same three fundamental steps but differ in the methods used to generate and deposit material.

Advantages and disadvantages of PVD:

PVD has several advantages including:

➤ Coatings formed by PVD have good strength and highly durable

➤ All types of inorganic materials and some types of organic materials can be used

➤ This process is environment friendly

However, PVD also has some disadvantages including:

➤ Problems with coating complex shapes

➤ High process cost and low output

➤ Some PVD methods operate at very high temperatures and vacuums, requiring special attention by operating personnel.

➤ Cooling systems are required to dissipate large heat loads.

4.11 Thermal Evaporation

Thermal evaporation is a physical vapor deposition technique that is generally operated under a high vacuum environment. This is usually used to produce high quality thin films at the nano level on almost all kind of substrates that include glass, metals, semiconductors, plastic, etc.

In a typical thermal evaporation experiment the material to be deposited is loaded into a container called as a crucible. The substrate is then placed just above the crucible. The crucible is then resistively heated by applying a large current. The material in the crucible starts to vaporize when they are heated; leaving the gaseous atoms to travel in straight lines until they reach the target and initiate the formation of a thin film. In order to improve the adhesion between the substrate and reacting species, the substrate are also heated in parallel. Here, the substrate temperature plays a significant role on the structural and optical properties of the material deposited. This method is usually followed to deposit materials on the substrates like plastic that cannot withstand high temperature treatment.

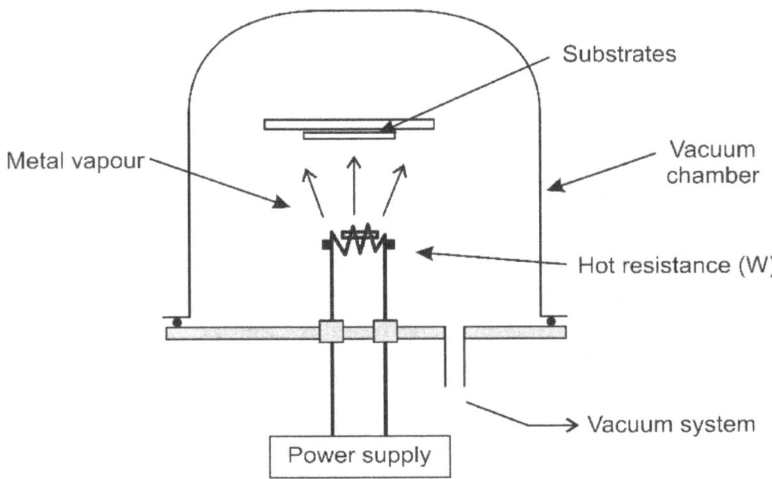

Fig. 4.9 Schematic diagram of thermal evaporation system [13]

4.12 Pulsed Laser Deposition (PLD)

Pulsed laser deposition (PLD) is considered to be one of the novel physical vapour deposition techniques in thin film deposition. In a typical PLD experiment, a high- power pulsed laser beam is allowed to strike the target material placed inside a vacuum chamber. The intensity of the incident beam vaporizes the target material, which gets deposited in the form of thin film on the substrate such as silicon. This process can occur in vacuum or in the presence of inert ambience, sometimes oxygen gases are also used as background, especially while depositing oxide materials. Such treatments are believed to assist in oxygenating the deposited films efficiently.

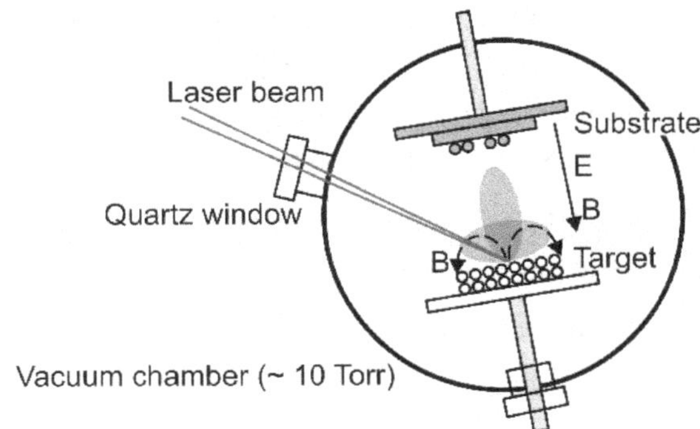

Fig. 4.10 Schematic illustration of a PLD system [14]

One of the unique features in PLD lies with the laser source, which is held outside the deposition chamber. This setup helps to establish a dynamic range of operating pressures for thin film deposition. Due to the highly reduced spot size of the laser source, the target area can also be prepared even under less than $1cm^2$. As a digital technique PLD has also displayed its potential to control the growth of thin films at atomic levels. Moreover, by controlling the deposition temperature and pressure it is also feasible to attain a variety of nanostructures and nanoparticles with unique functionalities. Some of the experimental factors that influence the crystallinity, uniformity, stoichiometry and growth kinetics involved in a PLD reaction have been listed below.

➢ Target material

➢ Distance from target to substrate

➢ Pressure in the chamber

➢ Intensity of the laser beam

➢ Surface temperature

➢ Ionization degree of the ablated material

➢ The nature of the substrate

4.13 Sputtering Technique

Sputtering is a physical vapor deposition process that is used to deposit thin films. In case of sputtering process, the atoms are ejected from a solid target material by bombarding with high energetic particles of a plasma or gas. The inert gases such as argon are commonly preferred as the sputtering gas in most of the experiments. The energetic ions are generated by the ionization of gas using DC, AC or RF input signal between the two electrodes - the cathode (target) and anode (substrate) of the system. The sputtering process consists of the following stages

➢ ionization of sputtering gas, usually an inert gas such as argon

➢ acceleration of ions towards the target

➢ collision between ions and atoms on the surface of target material

➢ ejection of atoms from target and

➢ Deposition of sputtered atoms onto the substrate.

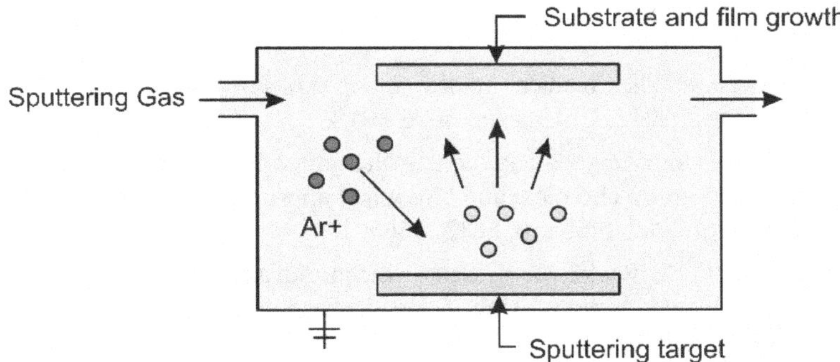

Fig. 4.11 Schematic illustration of a sputtering system [15]

Plasma generates a large number of positive ions between the electrodes. These ions are accelerated by plasma potential and bombard on the surface of target. The atoms will be knocked out from surface of the target material because the kinetic energy of incident particles is much higher than bond energy of solid. These sputtered and ejected atoms condense on the substrate to develop a thin film or nanowires with the help of catalysts. Sputtering is a versatile technique to deposit almost any type of material. Compared with other thin film technologies, sputtering is simple.

Sputtering yield:

Sputtering yield is defined as the number of atoms ejected from the surface per incident. It determines the deposition rate.

$$\text{Sputtering yield} = \frac{\text{No. of emitted particles}}{\text{No. of incident particles}}$$

Sputtering yield depends on the following factors

➤ *Mass of bombarding ions: The sputtering gases are usually opted by considering the atomic weight of the target. The atomic weights of the ions and the target atoms should be close.*

➤ *Energy of bombarding ions: The sputtering phenomenon takes place only while kinetic energy of the incoming particles is much higher than the conventional thermal energies.*

➤ *Direction of incidence of ions (incident angle): In oblique angle (45⁰ - 90⁰) there is a higher probability for sputtering, which occur closer to the surface*

➤ *Pressure: The required pressure for sustainable plasma is 10-1000 m Torr. The sputtering gas pressure can impact on the film deposition parameters such as deposition rate and composition of the film.*

References

1. Hai-Dong Yu et al, Chemical routes to top-down nanofabrication, Chemical SocietyReviews, 2013, Vol. 42, pp 6006-6018.

2. Rajagopalan Thiruvengadathan *et al,* Nanomaterial processing using self-assembly-bottom-up chemical and biological approaches, Reports on Progress in Physics, Vol. 76, Number 6, 2013, 066501.

3. G. W. Cave, C. L. Raston, J. L. Scott, Recent advances in solventless organic reactions: towards benign synthesis with remarkable versatility, Chemical Communications, 2001, pp2159-2169.

4. Yong X. Gan, Ahalapitiya, H. Jayatissa, Zhen Yu, Xi Chen, and Mingheng Li, Hydrothermal Synthesis of Nanomaterials, Journal of Nanomaterials Volume 2020, Article ID 8917013, pp 1-3.

   https://www.researchgate.net/figure/Stainless-steelautoclave_fig1_283153429

5. J Marques-Hueso et al, Photolithographic nanoseeding method for selective synthesis of metal-catalysed nanostructures, Nanotechnology, 2019, Vol.30 015302

   https://www.sciencedirect.com/topics/materials-science/optical-lithography

6. Amit Kumar, Nishtha Yadav, Monica Bhatt, Neeraj K Mishra, Pratibha Chaudhary and Rajeev Singh, Review Paper Sol-Gel Derived Nanomaterials and It's Applications, Research Journal of Chemical Sciences, ISSN 2231-606X, 2015, Vol. 5(12), pp 98-105.

   https://www.intechopen.com/chapters/64744

7. Gozde Ozaydin-Ince1, Anna Maria Coclite2 and Karen K Gleason2, CVD of polymeric thin films: applications in sensors, biotechnology, microelectronics/organic electronics, microfluidics, MEMS, composites and membranes, IOP Publishing Reports on Progress in Physics, 2012, Vol.75, 016501 (40pp).

8. M. Stueber H, Holleck H, Leiste K, Seemann S, Ulrich C. Ziebert, Concepts for the design of advanced nanoscale PVD multilayer protective thin films, Journal of Alloys and Compounds, 2009, Vol. 483, Issues 1–2, pp 321-333.

   https://en.wikipedia.org/wiki/Physical_vapor_deposition

   https://www.icmm.csic.es/fis/english/evaporacion_resistencia.html

   https://www.nature.com/articles/s41598-017-02284-0-PLD

   https://wiki2.org/en/Sputter_deposition

Chapter 5

Characterization Techniques

5.1 Introduction

Nanomaterial characterization is an important step for the development and adoption of nanotechnology for certain applications. The novel physical and chemical properties of these materials gave rise to a number of characterization techniques. Therefore, nanoparticles normally are characterized by studying their physical and chemical properties. They include composition, structure, size, morphology, surface area, optical properties, surface composition, oxidation state, electrochemistry etc., In view of this, for the complete characterization of a nanomaterial should not be limited to a single technique. This is because different types of measurements are needed for the complete characterization of a given nanomaterial. Therefore, in this chapter, the details of different characterization techniques such as X-ray diffraction (XRD), Transmission electron microscopy (TEM), Atomic force microscopy (AFM) and Scanning electron microscopy (SEM) etc., are discussed.

5.2 X-Ray Diffraction (XRD)

X- ray Diffraction (XRD) is one of the powerful, versatile, non-destructive techniques [1] used to characterize and analyze the nanoparticles. In addition to this, the other structural parameters such as lattice parameter, crystallite size, volume of unit cell, X-ray density and surface area of the material may be calculated. X-rays are the electromagnetic radiation with a very short wavelength about 1Å ($\lambda \approx 0.1$ Å) which is comparable with the inter-planer spacing of solids.

To determine the structure of a nanomaterial and thereby to ascertain the positions of its atoms in the lattice, a collimated beam of X-rays are directed toward the sample and is diffracted from the crystalline planes. The interaction between the X-rays and the material can be elucidated based on the Bragg's law, which is defined by an equation,

$$n\lambda = 2d\sin\theta \qquad\qquad .....(5.1)$$

where λ is the incident X-rays wavelength, n is an integer, d is inter-planar separation, and θ is Bragg angle or diffraction angle.

According to Bragg's principle, when X-rays incident on the sample, the lattice planes of the powder sample scatter the X- rays in all directions. If the total path difference between two diffracted rays is equal to $n\lambda$, they may interfere constructively and produce the X-ray diffraction pattern. The diffraction pattern can be used to identify the crystalline phases and their structural characteristics. The Fig. 5.3 shows the XRD pattern of nanocrystalline Cr-Co ferrite

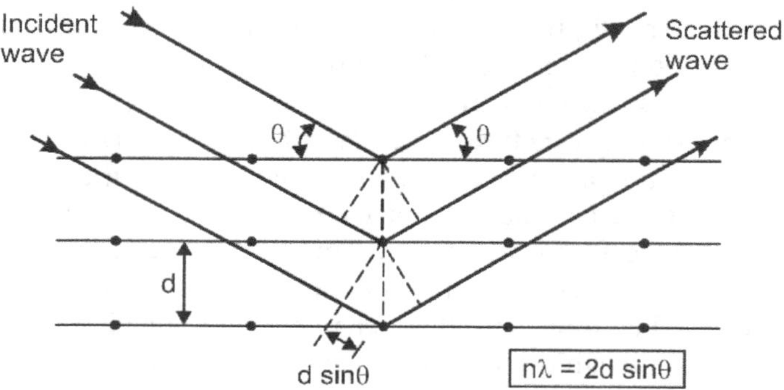

Fig. 5.1 X-ray diffraction on crystal planes

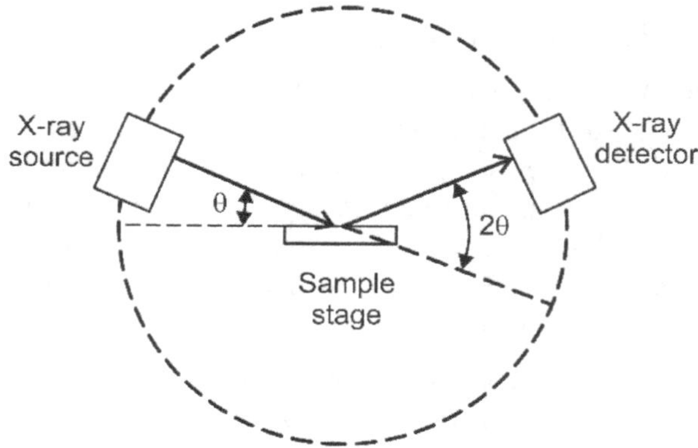

Fig. 5.2 Schematic diagram of X-ray diffractometer

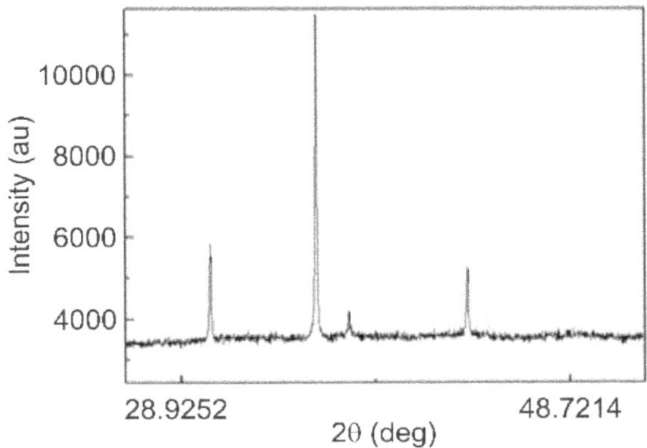

Fig. 5.3 XRD pattern of nanocrystalline Co- Cr ferrite

Lattice Parameter (a)

Lattice parameter (a) can measure by using the following formula

$$a = d_{hkl} \Big/ \left(\sqrt{h^2 + k^2 + l^2} \right)$$

where, d_{hkl} is the interplanar distance for hkl planes.

Unit Cell Volume (V)

Volume of the unit cell "V" can be calculate as

$$V = a^3$$

X-ray density (ρ_x)

X-Ray Density (ρ_x) can measure by using the following formula

$$\rho_x = 8M/Na^3$$

where, M is the molecular weight of the composition, N the Avogadro's number (6.023×10^{23} atom/mole).

Crystallite size (D)

The crystallite size (D) of a material may be defined by the well-known Scherer's Equation for the line broadening of the peak and is given by an equation,

$$D = k \, \lambda / \beta \cos \theta,$$

where λ is the wavelength of the X-ray; β, FWHM width of the diffraction peak; θ, diffraction angle; and k, constant. The crystallite size of nanomaterials is evaluated by calculating the Full Wave Half Maximum (FWHM) of the most intense peak from XRD pattern and by using the following Debye-Scherer equation,

$$D = 0.9\lambda \,/\, \beta\, \mathrm{Cos}\theta$$

where, D is the crystallite size, 0.9 is the symmetry constant, λ is the wavelength of incident x-ray, β is the full-width at half-maximum (FWHM) and θ *is* the diffraction angle. However, the accuracy of the technique used is always doubtful. The error in the values is very high and is as high as 10- 15%.

5.2   Fourier Transform Infrared Spectroscopy (FTIR)

Introduction

Infrared spectroscopy is the measurement of the interaction of infrared of the electromagnetic spectrum with matter by absorption, emission, or reflection. It is used to study and identify chemical substance or functional groups in a solid, liquid, or gaseous forms. Infrared Spectroscopy gives information about the rotational and vibrational modes of motion of a molecule. The Infrared spectroscopy in general provides a unique fingerprint, which is different from the absorption spectroscopy. Hence it is an important technique for characterization of materials in general and nanomaterials in particular. In fact, studies on the relations between structure and the electromagnetic response of nanoparticles are very useful in understanding their properties.

Electromagnetic radiations in which wavelengths lie in the range of micron to 1 nm are called as Infrared radiations. It has a little longer wavelength and lower frequency than that of visible light. The radiation was discovered by Sir William Herschel in 1800. Infrared region of the electromagnetic portion can be grouped into three different categories based on their wave numbers [2].

➢ Far IR (4 cm^{-1} ~ 400 cm^{-1})

➢ Mid IR (400 cm^{-1} ~ 4,000cm^{-1})

➢ Near IR (4,000 cm^{-1} ~ 14,000cm^{-1}).

IR Spectroscopy versus FTIR Spectroscopy

In the IR spectroscopy, the samples are irradiated with IR signals of different wavelengths and the process is continued for several hours. Later the data are analysed step by step taking the data of different times at a time. The entire process is laborious and takes several hours. FT-IR on the other hand, collects the spectral data of all wavelengths at a single time. In the process, the entire process can be completed in a quick time. Moreover, a continuous source generates IR light over a wide range of infrared wavelengths at a time. Additionally, in FT –IR spectroscopy the signal-to-noise ratio is very high. Therefore, FTIR is more advantageous than the IR spectroscopy.

Instrumentation

A typical spectrometer mainly comprises of components such as radiation source, optical path and monochromator, radiation detector and sample as shown in following Figure. These components are briefly described in Fig.5.4. below.

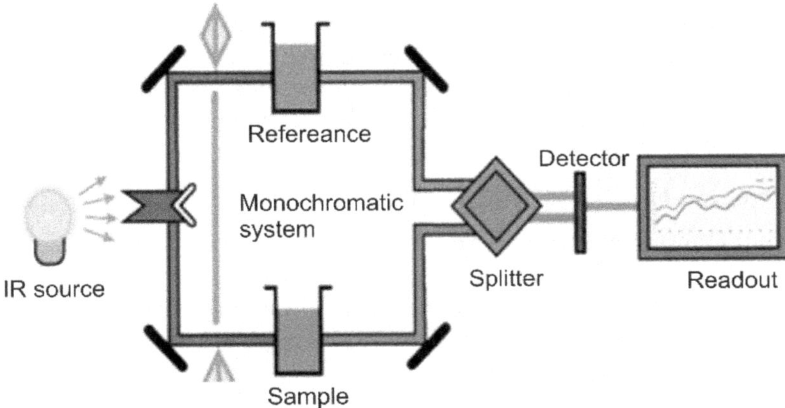

Fig. 5.4 Schematic diagram of FT-IR spectrometer [3]

Source: The IR source in this instrument is in the form of filament which is maintained at red- hot by an electric current.

Optical path and monochromatic: The IR beam from the source is focused on to the sample by the mirrors placed at different locations and the monochromator as shown in Fig.5.4.

Detector: Two different types of detectors are in common use. One of them is used for sensing the heating effect of the radiation while the other one is used for sensing the photoconductivity. In both the cases greater the effect (temperature or conductivity rise) at a given frequency, greater is the transmittance of the sample.

Sample: For the preparation of sample, KBr pellet method is normally used. In this method, the sample after finely grinding is mixed with pure and dry KBr powder. Later, the mixture is pressed in a hydraulic press to form a pellet. A major advantage of this method is that, KBr has no absorptions in the IR above 250 cm-1, so that an unimpeded spectrum of the compounds is obtained.

Working of FTIR Spectrophotometer

Infrared spectroscopy (IR-Spectroscopy) is a widely used technique for identifying the vibrational frequencies of chemical bonds in a given material. When a sample is exposed to infrared radiation, chemical bonds may absorb an amount of energy (hv) thereby reaching excited vibrational state. This may result in stretching the chemical bonds. The frequency of an absorption band is proportional to the energy difference between the vibrational state and the excited state. In the interferometer, an infrared beam from the source strikes a beam splitter and is separated into two beams. These two beams are reflected by the mirrors and are later recombined. The optical path difference between the two beams gives rise to constructive and destructive interference and the resulting interferogram is recorded and is interpreted as Fourier Transformation.

Advantages of FTIR:

Fourier Transform Infrared Spectrometers produces routine spectra in shorter time periods than those from conventional prism or grating systems. Superior resolution and sensitivity, quick sample measurement, versatile spectra processing, absolute wavelength accuracy and complete automation of complex analysis are main reasons make infrared spectroscopy very useful in material characterization.

5.3   Scanning Electron Microscope (SEM)

Among various characterization techniques, electronic microscopy plays a vital role in understanding the nature of nanoparticles. Researchers in biology, chemistry and physics apply this technique to observe structures that may be as small as 1 nanometer. The SEM may be employed to study organelles and DNA in cells, synthetical polymeres, and coatings on microchips. The SEM has several benefits over optical microscopes. Moreover, it has higher resolution, through which nearly spaced specimen grains can be extremely magnified. These features render SEM one of the most useful characterization techniques. SEM is essentially a high resolution electron microscope which uses a collimated beam of electrons instead of light to produce images of the specimen [4]. SEM provides topographical and elemental information of the sample at magnifications of 10 times to 30×10^4.

Working principle of SEM

A collimated electron beam impinges on a thin surface layer of a specimen and leads to the generation of backscattered electrons of high energy, secondary electrons of low energy and characteristic X- rays etc. These signals can be detected by suitable detectors and sent to the computer screen, which produces final image. The characteristic X-rays generated are used for the identification and estimation of different elements present in the specimen by energy dispersive spectrometer (EDS).   In order to avoid the oxidation and contamination of filament (tungsten tip) as well as to reduce the collisions between air molecules and electron, filament and sample are kept in a vacuum chamber. Usually, vacuum of the order of 10^{-5} torr or better is necessary for a normal functioning of SEM. The schematic representation of SEM is shown in Fig. 5.5.

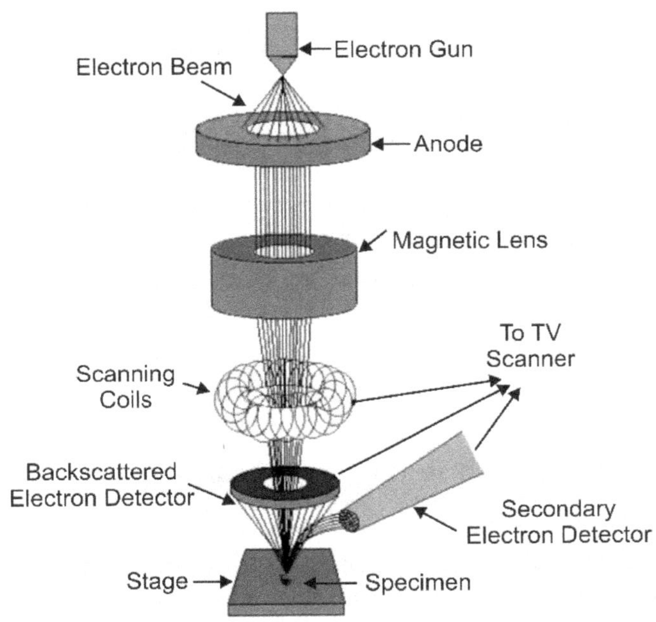

Fig. 5.5 Schematic Block diagram of SEM [5]

When the high energy electron beam strikes the sample, various interactions as shown in Fig.5.6 may be possible.   They include Back scattered electrons, secondary electrons, X-ray photons etc.as shown in Fig. 5.6. For final imaging, normally both secondary and backscattered electrons are used.

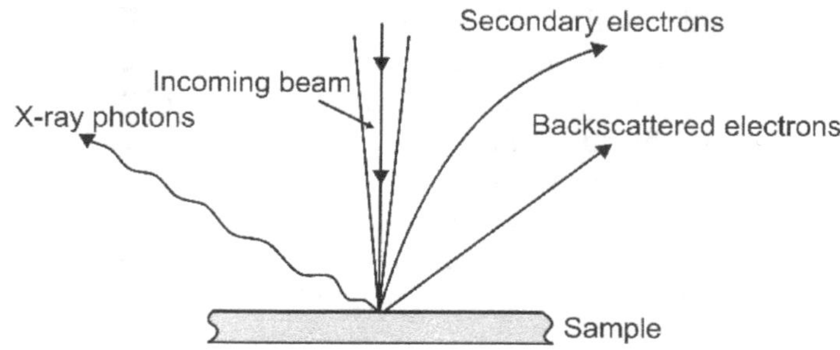

Fig. 5.6 Schematic diagram showing the electron-specimen interaction

Advantages of SEM:

➤ SEM gives the different aspects of crystallite information such as atomic arrangement in the specimen and their degree of order, topography, morphology, the elements and compounds the sample is composed of and their relative ratios, etc.

➤ Wide range of applications such as, topographical 3D image of the sample.

➤ Sample preparation is easy when compared with that of TEM

➤ Fast and easy operation

➤ Interpretation of images is easy

➤ Versatile information obtained from different detectors

Disadvantages of SEM:

➤ SEMs are expensive, large and must be housed in an area free from any possible electric, magnetic or vibration interference.

➤ Maintenance cost is high

➤ Special training is required to operate an SEM

➤ The preparation of samples can result in artifacts.

➤ Resolution is limited to tens of nm

➤ Details on the atomic level cannot be observed

➤ There is risk of radiation exposure in it due to scattered electrons

5.4   Transmission Electron Microscope (TEM)

Transmission Electron Microscope (TEM) is a powerful imaging tool with high resolution to examine nanoparticles that are too small for observation with conventional microscopes and it usually provides more detailed geometrical features than those obtained with SEM. TEM pattern is exactly similar to the optical microscope except that a collimated beam of electrons is used in lieu of light rays to "see through" the test sample.

TEM was designed by Max Knoll and Ernst Ruska in 1931. TEM studies provide information regarding the crystal structure, crystal quality, and grain size, particle shapes as well as their size and degree of agglomeration. In TEM, electrons are emitted by thermionic emission from a hot tungsten filament and are focused by the magnetic lenses. The interior of TEM must be under extremely high vacuum (less than 10^{-7}atm) to inhibit scattering of electrons by air molecules. TEM is a valuable tool in material science, nanoscience and biological science as the instrument gives a very high magnifications of the samples. One may obtain a resolution of 1Å and can see very small objects (<1nm) [4].

Working principle of TEM

The detailed schematic representation of a TEM is shown in Fig. 5.7. Each component /part is labeled, and their functions are briefly explained below. All the lenses which are used in this instrument are electrostatic /magnetostatic because the electrons are used to form the image instead of photons.

> A field emission gun at the top of the microscope produces a stream of electrons.

> A condenser lens produces electrons into a narrow coherent beam.

> The condenser aperture eliminates high angle electrons.

> The accelerated electron beam incidents on the test sample and some portion of it is transmitted.

> The transmitted portion of electron beam is focused by an objective lens.

> Selected area aperture facilitates the observer to examine the diffraction by an ordered arrangement of atoms in the specimen.

> Intermediate and projector lenses magnify the image.

> The beam strikes the phosphor screen to form an image.

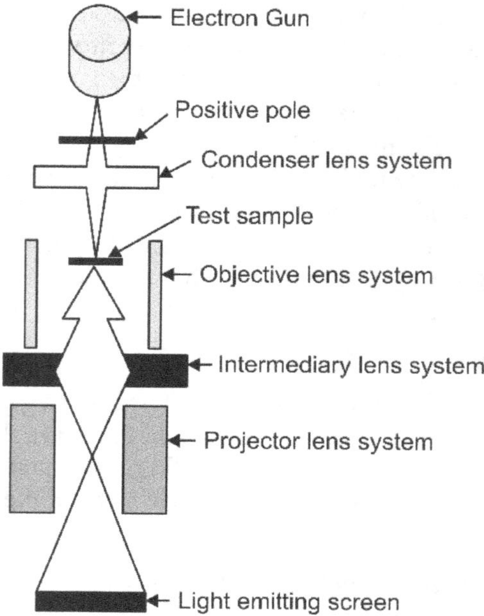

Fig. 5.7 Block diagram of Transmission Electron Microscope

In TEM, an electron beam with an energy of 100keV or more (up to 3MeV) is allowed to go through a thin specimen. TEM offers special resolution down to an Angstrom and high magnification up to 10^6. The high resolution of TEM might be due to small wavelength of electrons, given by de Broglie relationship:

$$\lambda = h / (2mqV)^{1/2}$$

where h is Planck's constant, m and q are mass, and charge of the electron and V is applied potential difference through which electrons are accelerated. The higher is the operating voltage of a TEM, the greater its resolution.

Advantages of TEM:

➢ Transmission electron microscopes provide magnification of one million times.

➢ It produces highest resolution images.

➢ It has wide range of applications in educational, scientific and industrial fields.

➢ TEM can provide information on element and compound structure it forms detailed and high quality image of the samples.

➤ Easy operation

➤ Yield information of surface features, shape, size and structure.

Disadvantages of TEM:

➤ TEMs are large and very expensive

➤ Extensive sample preparation is required to produce a sample thin enough to be electron transparent.   Biggest problem with this instrument is the sample preparation.

➤ TEM analysis is a relatively time consuming process and it takes more time for each study.

➤ During the preparation of the sample, the structure of the sample may change occasionally. There is also the possibility of the sample being damaged by the electron beam.

➤ Complicated contrast mechanisms which makes image interpretation difficult.

5.5   Scanning Tunneling Microscope (STM)

The scanning tunneling microscope (STM) developed by Dr. Gerd Binnig and Heinrich Rohrer in 1981, is the first instrument capable of directly obtaining three-dimensional (3D) images of solid surfaces with atomic resolution. Their discovery opened a new era for surface science, and their impressive achievement was recognized with the award of the Nobel Prize for Physics in 1986.   STM functions on a principle of operation on tunneling. This is possible because wave like properties of electrons permit them to "tunnel" beyond the surface of a solid.   A small voltage is applied between the probe tip and the surface, causing electrons to tunnel across the gap. As the probe is scanned over the surface, it registers variations in tunneling current. This information after processing may provide a topographical image of the surface.

Components of STM:

The main components of STM are a) Scanning tip b) Piezoelectric scanner c) Distance control d) Scanning unit e) Vibration isolation system f) Data processing unit

1. **Scanning Tip:** STM tips are usually made up of tungsten metal or a platinum-iridium alloy.   Scanning tip is the most important component of the STM.

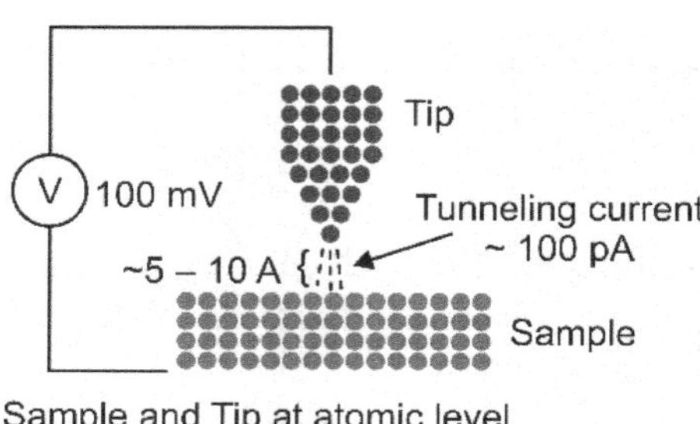

Sample and Tip at atomic level

STM tip under
$10^3 \times 10^5 \times$ magnification

Fig. 5.8 STM tip   **Fig. 5.9** Magnified image of STM

2. **Piezoelectric Scanner:** The scanner tip is attached to a piezoelectric tube scanner and works under the phenomenon piezoelectric effect. Under this effect, the material changes its length when put in an electrical voltage. By adjusting the voltage on the piezoelectric element, the distance between the tip and the surface can be adjusted. Piezoelectric crystals expand or contract slightly depending on the voltage applied. In fact, based on this principle the horizontal position and *vertical positions* of the scanning tip are also controlled.

3. **Distance Control and Scanning unit:** Controlling the position of the scanning unit is normally carried out using piezoelectric method.

4. **Data Processing Unit (Computer):** The computer records the tunneling current and controls the voltage to the piezoelectric tubes to produce a 3-dimensional map of the sample surface.

5. **Vibration Isolation System:** Due to extremely high sensitivity of tunneling current between tip and sample surface height, it is absolutely necessary to reduce inner vibrations and to isolate the system from external vibrations. Damping can be achieved by Pneumatic methods, Eddy current system etc.

Working Principle of STM

First, the tip is brought close to the sample by coarse control arrangement system. Subsequently, the fine control of the tip is to be operated with piezoelectric method. Now, a bias voltage is applied between the tip and the sample surface allowing the. Electrons to tunnel through the vacuum between them. The resulting tunneling current (I_t) is exponentially dependent on the distance between tip and sample. If the distance increases, the tunneling current (I_t) decreases exponentially as per the relation,

$$I_t = Ve^{-kd}$$

V: potential difference between conductors.

k: constant which depends on conductors composition

d: distance between the lowest (nearest to the sample) atom on the tip and the highest (nearest) atom on the sample.

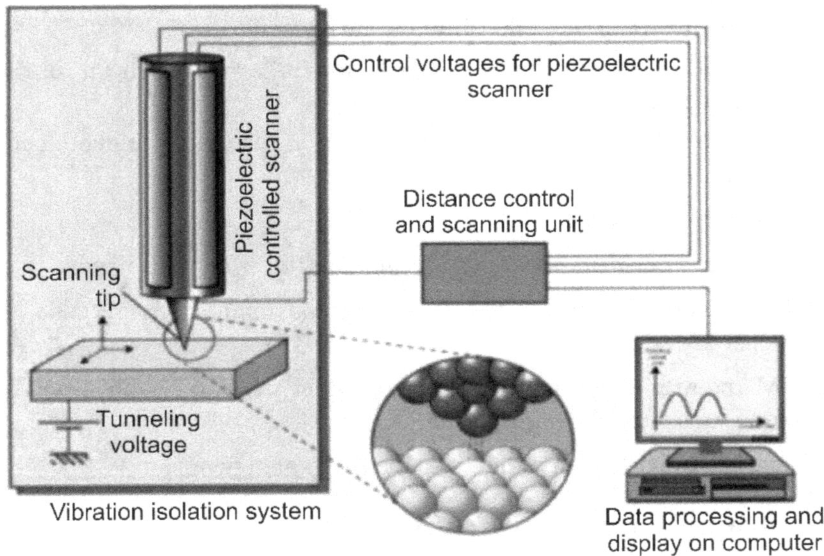

Fig. 5.10 Schematic diagram of basic STM set-up [6]

For tunneling to take place, both the sample and the tip are to be conductors or semiconductors. However, STM cannot work with insulating materials. In this instrument, the convention of the polarity of bias voltage is used. According to the convention of polarity, when the tip is grounded and bias voltage (V) is applied to the sample, tunneling current is passed. For example when V = 0, no tunneling current is observed. If V > 0, the electrons are tunneling from the occupied states of the tip into the empty states of the sample. If V < 0, the electrons are tunneling from the occupied states of the sample into the empty states of the tip. If the tip is moved across the sample in the x-y plane, the changes in surface height and density of states change in current. This change in current with respect to position can be measured and used to create an STM image.

***Advantages of STM*:**

> STM gives three dimensional profile of a sample surface, which allows researchers to examine a multitude of characteristics, including roughness, surface defects and molecule size.

> STMs are versatile and these instruments can be used in ultra high vacuum, air, water and other liquids and gasses.

> They will activate in temperatures as low as zero Kelvin up to a few hundred degrees Celsius.

> Lateral Resolution of 0.1 nm and 0.01 nm of resolution in depth can be achieved.

***Disadvantages of STM*:**

There are very few disadvantages of scanning tunneling microscope

> STM is very expensive. The operation and maintenance is also very expensive

> STM requires very clean surface, excellent vibration control while in operation, sharp and single atom tip.

> The operation of STM requires specially trained people.

5.6   Atomic Force Microscopy (AFM)

Atomic Force Microscope (AFM) is one type of scanning probe microscope, which has an ability to create three dimensional micrograph of sample surface with resolution down to the atomic level. It is therefore possible to extract quantitative data such as the concentration of particles, particle size, particle size distribution and surface roughness from the image. The AFM is one of the foremost tools for imaging, measuring and manipulating matter at the nanoscale. Atomic force microscopy is currently applied to various environments (air, liquid, vacuum) and types of materials such as metal semiconductors, soft biological samples, conductive and non-conductive materials. Materials investigated by the AFM are thin and thick films, ceramics, composites, glasses, metals, polymers and semiconductors.

Working Principle of AFM

Atomic Force Microscopy (AFM) measures the interaction force between the tip and the surface [7]. The interaction force normally depends on the nature of the sample, the probe tip and the distance between them. In fact, AFM provides a 3D profile of the surface on a nanoscale. The probe is supported on a flexible cantilever. The AFM tip "gently" touches the surface and records the even a small force between the probe and the surface.

The image is produced by dragging a vibrating cantilever with a Si_3N_4 tip across the surface. When a sample brought close to a vibrating cantilever, a weak interaction force exists between the tip and the film surface that cause the tip to deflect from its equilibrium position according to Hooke's law which states that,

$$F = -kx$$

where, F = force, k = spring constant and x = cantilever deflection

The motion of the probe across the surface is controlled using feedback loop and piezoelectric scanners. The deflection of the probe is measured by a "beam bounce" method. In this method first a laser beam is reflected on to the back side of the cantilever as shown in Fig. 5.11. From the cantilever, the laser beam is focused to a photo diode. In this arrangement even a small deflection in the cantilever is reflected in the detector arrangement. This detector measures bending of cantilever during the movement of tip. The measured cantilever deflections are used to generate a map of the surface topography.

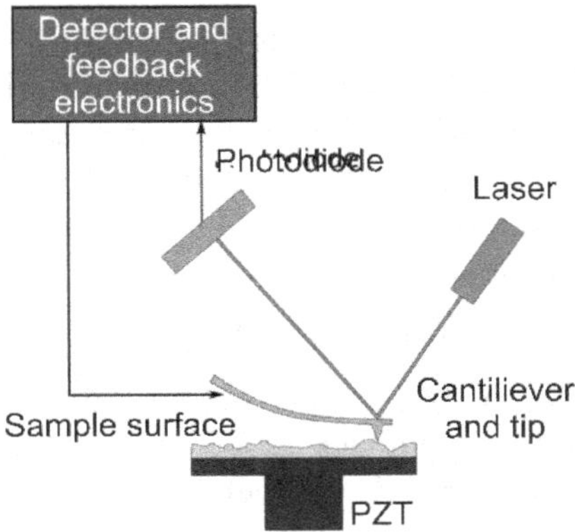

Fig. 5.11 Schematic diagram explaining principle of AFM [8]

The common AFM Modes of Imaging

The most common AFM modes of imaging (a) contact mode (b) tapping mode and (c) noncontact mode are described below:

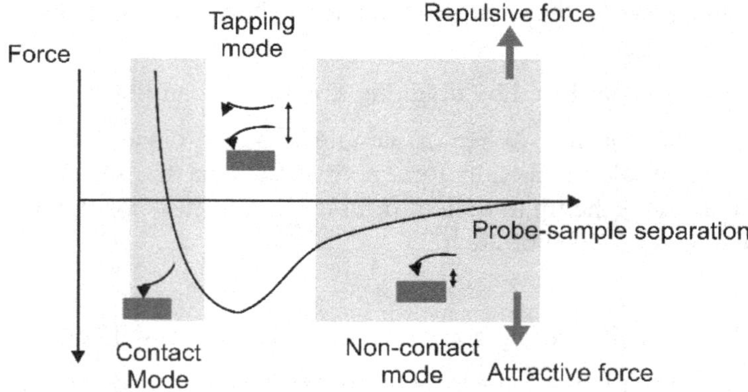

Fig. 5.12 Force Vs distance (tip to sample) curve

a) Contact Mode AFM (repulsive VdW)

When the spring constant of cantilever is less than that of surface, cantilever bends and the force on the tip is repulsive. By maintaining a constant cantilever deflection, the force between the probe and the sample remains constant and an image of the surface is obtained.

b) Tapping mode (Intermittent mode)

The image obtained in this process is similar to the contact mode. The only difference in this mode is that the cantilever oscillates at its resonant frequency. The probe lightly "taps" on the sample surface during scanning, contacting the surface at the bottom of its swing. By maintaining constant oscillation a constant tip-sample interaction is maintained and an image of the surface is obtained.

c) Non-contact mode (attractive VdW)

In this method, the probe does not contact the sample surface and as such the method is called non-contact mode method. However, the probe oscillates above the adsorbed fluid layer on the surface during scanning. Using a feedback loop to monitor changes in the amplitude due to attractive VdW forces the surface topography can be measured.

Advantages of AFM:

> The AFM provides a three-dimensional surface profile

> Determine the roughness of a sample surface

> Easy to prepare samples for observation, samples viewed by AFM do not require any special treatments (such as metal/carbon coatings)

> It can be used in vacuums, air, and liquids.

> It can be used to study living and nonliving elements

- ➢ Image non-conducting surfaces such as proteins and DNA
- ➢ Study the dynamic behavior of living and fixed cells
- ➢ Any sample like ceramic materials, polymers, glasses, human cells or individual molecules of DNA, dispersion of metallic nanoparticles can be imaged

Disadvantages of AFM:

- ➢ It can only scan a single nanosized image at a time of about 150×150nm.
- ➢ Limited scanning speed.
- ➢ Limited vertical range
- ➢ Imaging depends on tip shape
- ➢ The tip and the sample can be damaged during detection.

References

1. Andrei A. Bunaciu, Elena Gabriela Udristioiu and Hassan Y. AboulEnein, X-Ray Diffraction: Instrumentation and Applications, Critical Reviews in Analytical Chemistry (2015) 45, 289–299.

2. Shahid Ali Khan et al, Fourier Transform Infrared Spectroscopy: Fundamentals and Application in Functional Groups and Nanomaterials Characterization, Chapter 9, Handbook of Materials Characterization, Springer International Publishing A G, part of Springer Nature, 2018, pp.317-344.

3. https://microbenotes.com/infrared-ir-spectroscopy.

4. M. Kannan, Scanning Electron Microscopy: Principle, Components and Applications, In book: A Textbook on Fundamentals and Applications of Nanotechnology, publisher: Daya Publishing House, A Division of Astral International Pvt. Ltd. New Delhi, 2018, pp.81-101.

5. https://www.researchgate.net/figure/Shows-the-basic-block-diagramof-a Scanning-Electron-Microscope_fig3_254707658.

6. http://www.hk-phy.org/atomic_world/stm/stm03_e.html.

7. Julian Chen, Introduction to Scanning Tunneling Microscopy: Second Edition, American Journal of Physics, 1994, 62(6).

8. https://www.sciencedirect.com/topics/physics-andastronomy/atomic-force-microscopy.

Chapter 6

Properties of Nanomaterials

6.1 Introduction

The properties of nanomaterials are significantly different from those of micrometer and millimeter size or bulk materials. This may be mainly attributed to the fact that (i) these materials are having their large number of atoms on their surface (ii) high surface energy; (iii) confinement of large number of atoms and (iv) reduced number of imperfections, which do not exist in their bulk counterparts or micrometer size materials[1]. In view of this surface properties and hence the surface in nanomaterials is more interesting and important [2, 3]. Although, some of the peculiar properties are already known to the scientific community, a lot more is there to be discovered. The nanometre feature sizes of nanomaterials have spatial confinement effect on the materials, which bring the quantum effects. The quantum confinement has profound effects on the properties of nanomaterials. The energy band structure and charge carrier density in the nanomaterials are quite different from their bulk counterpart and this in turn will modify the electronic and optical properties of the materials. Another important parameter to be considered in these materials is the reduced imperfections. Infact, reduced imperfections, play an important role in determining some of the physical properties of nanomaterials. Further, in nanostructures and nanomaterials another important parameter viz., self-purification process plays an important role a vital role in moving the impurities and intrinsic material defects to the surface upon annealing. This phenomenon indirectly influences the mechanical properties of nanomaterials [4]. Therefore, nanoparticles display properties that are different from those of the individual atoms or bulk materials. Some of the physical, chemical, electrical, optical, magnetic, thermal and mechanical properties of nanomaterials are discussed in this chapter.

6.2   Physical Properties

Large Surface Area

In both the bulk materials or in nanoscale materials **physical and chemical properties depend on a lot on its surface phenomena.** For example, surface performs numerous functions: they include keeping things in or out; allowing the flow of a material or energy across an interface; initiate or terminate a chemical reaction.

In order to understand the concept of large surface area, let us take a bulk material and subdivided into an ensemble of individual pieces of nanomaterials. In the process, although the total volume remains the same, the **collective surface area goes up considerably.** This is schematically shown in following figure.

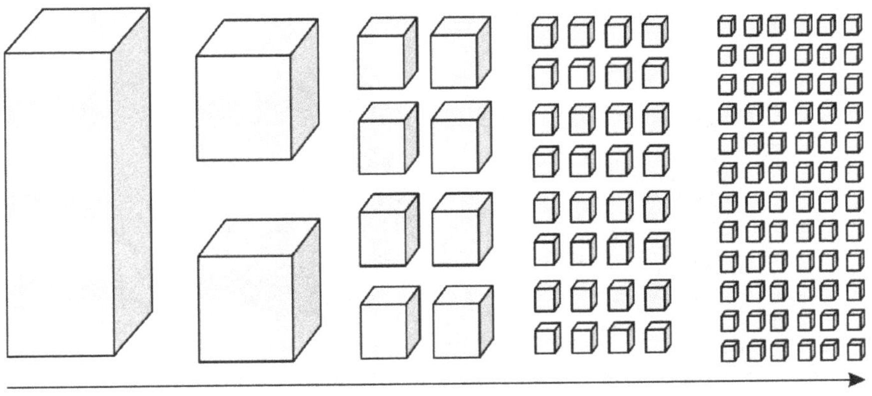

Fig. 6.1 Schematic drawing showing how surface-to-volume increases as size is decreased

The consequence is that the surface area of the nanomaterial is increased when compared with its bulk material counterpart. In order to understand the entire concept of surface area may be understood and explained by taking an example of 1 sq meter cube. How would the total surface area increases if a cube of 1 m^3 were progressively cut into smaller and smaller cubes, until it is formed of 1nm^3 cubes? Results are summarized in Table 6.1.

Table 6.1 Surface area of a material, which is cut into small cubes

Size of cube side	Number of cubes	Total Surface Area
1 m	1	6 m^2
0.1 m	1000	60 m^2
0.01 m = 1cm	10^6 = 1 million	600 m^2
0.001 m = 1mm	10^9 = 1 billion	6000 m^2
10^{-9} m = 1 nm	10^{27}	6×10^9 = 6000 Km^2

For any material, the surface area depends on the its shape. A sphere and a cube having the same volume are taken as an example to explain the concept of surface area in nanoscience. The cube has a larger surface area than the sphere. Therefore, one may conclude that in nanoscience not only the size, but also its shape is important.

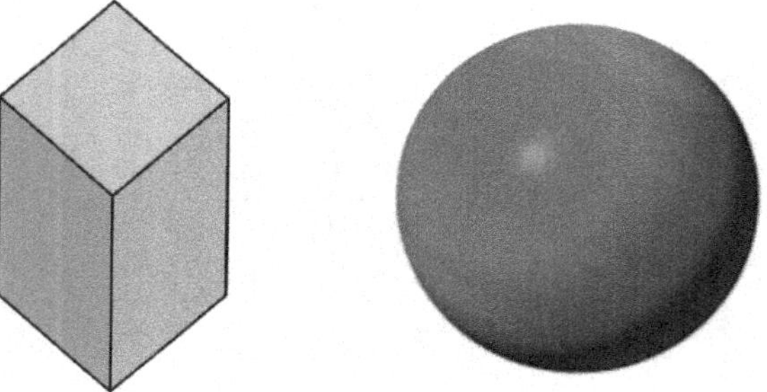

If side of a cube is taken as equal to 1 m, Surface area of sphere is 4 πr^2 the surface area of a cube is 6sq meters

The following figure illustrates this concept clearly in nanomaterial whose surface area is determined not only by size but also by shape.

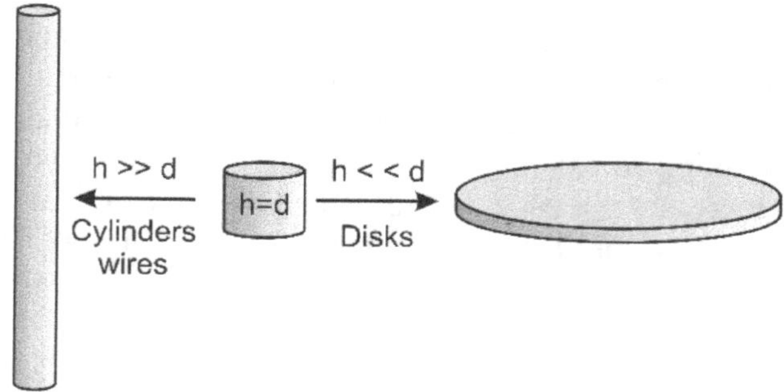

Fig. 6.2 The ratio between h and d determines whether a shape is like a wire or a disc.

Melting point of Nanoparticles

In bulk materials, since surface-to- volume ratio is very small, surface effects can be disregarded. On the other hand, in the case of nanomaterials, for which the ratio of surface-to- volume is large and the curvature of the surface is usually very pronounced. Consequently, for nanomaterials the melting temperature is size-dependent. Nanoparticles of metals, semiconductors and molecular crystals are all found to have lower melting temperatures as compared with their bulk forms, when the particle size decreases below 100nm. Melting-point depression normally means the reduction of melting point of a material with reduction of its size. This phenomenon is very prominent in nanoscale materials which melt at temperatures hundreds of degrees lower than their bulk material counter parts. Melting-point depression is clear in nanowires, nanotubes and nanoparticles, which all melt at lower temperatures than their bulk material counterpart.

The melting point of a material is directly correlates with the bond strength. Nanoscale materials have a much larger surface-to-volume ratio than bulk materials, because of this in case of nanomaterial a larger fraction of the atoms are present at the surface. These surface atoms can be easily removed than those in bulk atoms, so the total energy needed to overcome the intermolecular forces that hold the atom "fixed" is less and thus the melting point is lower [5]. As the particle size decreases, the melting temperature also decreases. For example, melting point of bulk gold is of 1337K and decreasing rapidly for nanoparticles with sizes below 5nm as shown in following figure [6].

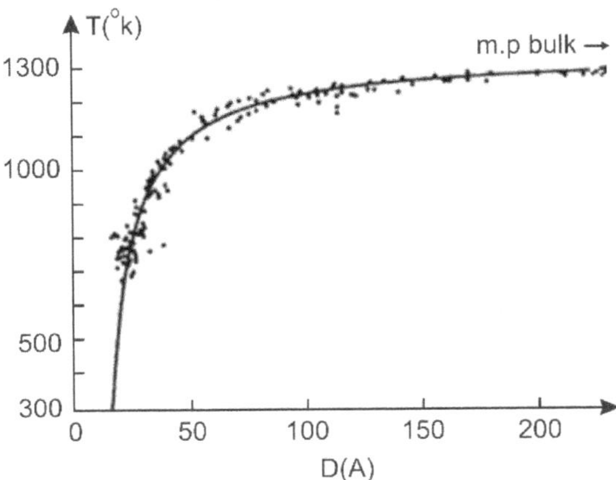

Fig. 6.3 Variation of melting point as a function of particle diameter of gold

Reduced Imperfections in Nanomaterials

Nanomaterials favor a self- purification process in which the impurities and defects will move to near the surface upon thermal annealing. Self purification is the intrinsic thermodynamic property of nanostructures and nanomaterials. Reduction of impurities in nanomaterials   is an important phenomenon and will help the materials in improving their performance [4].

Most metals are made up of grain boundaries and are responsible for slowing down the propagation of defects [7]. If these grains are in nanosize, the grain boundary within the material greatly increases, enhancing its strength. The strength of nanomaterials is one or two orders of magnitude higher than that of single crystals in the bulk form. Thus the improvement of mechanical strength in nanomaterials could be due to the reduction of defects.

Diffusion

Diffusion is the movement of particles of a solid from the place of high concentration to a place of low concentration resulting in uniform distribution. Diffusion is a passive process, wherein no assistance from outside energy   is needed. The process is continued until equilibrium is reached.

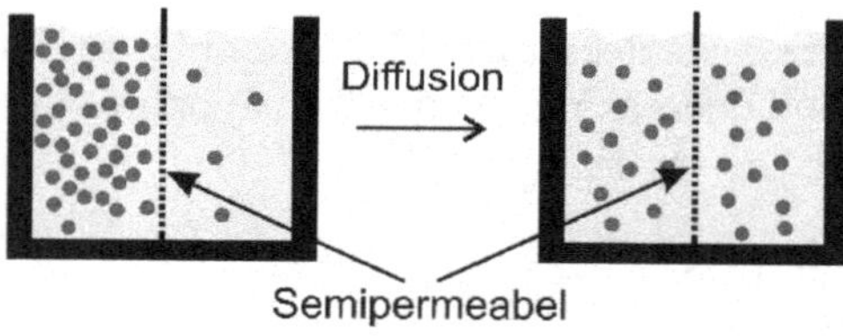

Fig. 6.4 Diffusion process

Rate of diffusion is independent of time. Further, the rate of diffusion is proportional to concentration gradient (dC/ dx).

Fick's first law of diffusion is defined as,

Molar flux change, $J = D$ (dC/ dx),

where D is the diffusion coefficient

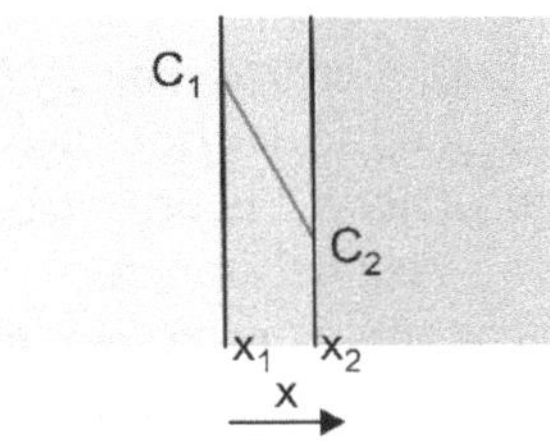

Fig. 6.5 Explanation of First Ficks Law

At any given temperature, the diffusion of a smaller particles are always larger than those of a bigger sized molecules. This is related to both the mass of the molecule and its surface area. A bulk material will diffuse slowly, while nanoparticles with a larger surface area will diffuse more quickly. In nanomaterials self-diffusion, solute diffusion and solute solubility are very high and the higher values may be attributed to defective atomic coordination at the grain boundaries [8]. The reduced coordination number of the surface atoms greatly increases the surface energy so that atom diffusion occurs at relatively lower temperatures.

Nanomaterials are composed of large number of grains and grain boundaries. Grain boundaries and dislocations in nanomaterials provide high density of short circuit diffusion paths compared with the bulk materials. Thus nanomaterials are expected to exhibit enhanced self diffusivity when compared with the single crystals or bulk materials with the same composition.

For example, the diffusivity in nanocrystalline copper is about 20 times higher than in bulk materials.

Table 6.2 Diffusivities in nanomaterials (in m^2/s at 300K)

S.No.	Name of the material	Bulk material	Nanomaterials
1.	$^{67}Cu/^{64}Cu$	10^{-39}	10^{-19}
2.	$^{107}Ag/^{64}CU$	10^{-39}	$2*10^{-19}$

6.3 Chemical Properties

Enhanced Chemical Reactivity

The chemical groups on the surface of a material determine its properties. For example, properties such as catalysis, electrical resistivity, adhesion, gas storage, chemical reactivity etc., depend on the nature of the surface. It is well- known that Nanomaterials have a large number of surface atoms. This has a strong influence on several surface reactions such as catalysis, detection reactions, physical adsorption etc., Surface atoms and molecules are unstable; they have high surface energy and enhanced reactivity. According to fundamental chemical principle, "systems of high energy will strive to attain a state of lower energy (for the sake of stability) by whatever means possible". As mentioned above, nanomaterials have a very large fraction of their atoms and molecules on their surface. Hence, nanomaterials are highly unstable and they participate actively in all possible chemical reactions to attain stability. Therefore we can conclude that the nanomaterials have enhanced chemical reactivity when compared with that of bulk materials. This means that materials that are inert in their bulk form are reactive when produced in their nanoparticle form [9].

Ex 1: A good example is the gold. Bulk gold is stable, non-toxic, resistant to oxidation and chemically inert, while the nanogold is chemically very active and is having several interesting chemical properties.

Ex 2: Another important example is, the absorption of Hydrogen in nanocrystalline Pd is 10 - 100 times larger than in Pd single crystals (Bulk materials). This shows that nanoparticles may be very useful in hydrogen storage devices in metals.

Enhanced Solid Solubility

Solubility plays an important role in drug disposition, since maximum rate of the drug transport across a biological membrane, the main pathway of drug absorption, is the product of permeability and solubility. Solubility plays an important role in drug disposition, since maximum rate of the drug transport across a biological membrane, the main pathway of drug absorption, is the product of permeability and solubility.

At a given temperature, the solubility of nanoparticles is more rapid than that their bulk counter parts and the the behavior may be attributed to both the mass of the molecule and their surface area. Therefore, one may conclude that a bulk material dissolves slowly, while nanoparticles with a larger surface area dissolves faster manner.

Ex 1: A very simple example how nanoscience can impact on the properties of the materials is granular sugar and caster sugar. Caster sugar is finer, stickier (more surface absorption) and in water dissolves faster.

Ex 2: Similar effects were also reported in the case of dissolution of Bismuth in Cu. The solubility of Bismuth in bulk Cu is less than at 10^{-4} at 100°C. On the other hand in the case of nanocrystalline Cu, the dissolution of Bi is about 4%, indicating that the solubility has enhanced to more than 10,000 times.

We know that the possibilities of improving dissolution is based on the surface area available for dissolution. Therefore, by decreasing the particle size of a solid compound [10], the solubility in nanomaterials enhances. Thus, solubility plays an important role in drug transportation as drug transportation across a biological membrane is the product of permeability and solubility.

6.4 Electrical and Electronic Properties

The influence nanosize on electrical conductivity of nanomaterials is complex, as both these properties depend on different mechanisms and they may be grouped into categories

i)   surface scattering including grain boundary scattering

ii)   quantized conduction

iii)   coulomb charging and tunneling

iv)   widening and discrete of band gap

v)   change of microstructures.

In addition, the reduced impurity and structural defects along with reduced dislocations, would affect the electrical conductivity of nanomaterials.

Electrical Resistivity

Nanosize normally influences all the properties of materials including the electrical resisitivity. When the size is reduced, it will have two different effects on electrical resistivity. Firstly, the nanosize results in the reduction of defects, resulting in decreasing due to defect scattering thereby reducing the resistivity of a material. However, the defect scattering makes a minor contribution especially in metals and thus the reduction of defects has a very small influence on the electrical resistivity, mostly unnoticed experimentally. The other factor influencing the resistivity is due to surface scattering and plays an important role in determining the total electrical resistivity of nanosized materials [11]. If the mean free path of electron due to the surface scattering is the smallest, then it will dominate the total electrical resistivity. In nanowires and thin films, the surface scattering of electrons results in reduction of electrical conductivity. When the critical dimension of thin films and nanowires is smaller than the electron mean-free path, the motion of electrons will be interrupted due to collisions with the surface resulting in decreasing electrical conductivity.

Variation of Nanosize with Density of States

The overall behavior of bulk crystalline materials changes when the dimensions are reduced to the nanoscale. For example, in a 0-D nanomaterials, where all the three dimensions are at the nanoscale, the electron is confined in all the three 3-Directions. For 1-D nanomaterials, electron confinement occurs in 2-D, whereas in a 2-D nanomaterials, the conduction electrons is confined across the thickness only. As more number of the dimension is confined, more discrete energy levels can be found, which in turn results in the electron movement is strongly confined in a given dimension [12].

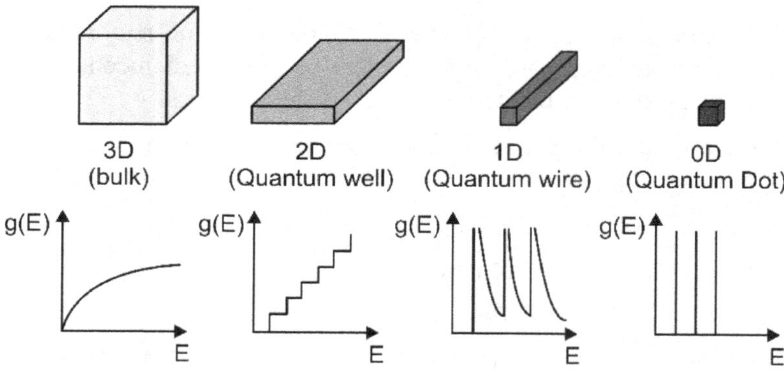

Fig. 6.6 Variation of density of states with dimensionality

As shown in above figure, a reduction in dimensions below the electron wavelength, may result in developing the discrete band gap. Such a change generally would also result in a reduced electrical conductivity. In fact, some metal nanowires may undergo a transition and may become semiconducting as their diameters are reduced below certain values, and semiconductor nanowires may become insulators. Therefore one may conclude that the electrical properties of nanomaterials strongly depends on particle dimensions.

Nano size versus Moore's law

Moore's law states that the number of transistors per square inch in an integrated circuit, doubles about every two years. Subsequently, 24 months period has been reduced to 18 months. The observation is named after Gordon Moore, the co-founder co-founder of Intel, whose 1965 research paper described a doubling every year in the number of components per integrated circuit. Infact, advancements in digital electronics are strongly linked to Moore's law:

This exponential change is one of the reasons why nanoscience has become important and relevant today. Nanoscience and technology are today at the same stage as information technology was in the 1960s and Biotechnology was in the 1980s. It is predicted that in the years to come, this subject may witness a similar type of exponential growth as the other two witnessed earlier.

today at the same stage as information technology was in the 1960s and Biotechnology was in the 1980s. It is predicted that in the years to come, this subject may witness a similar type of exponential growth as the other two witnessed earlier.

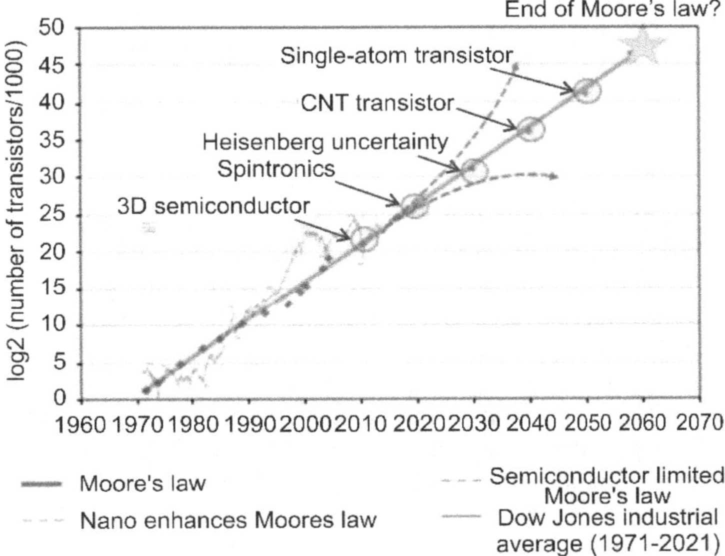

Fig. 6.7 Moore's Law history, future trends

Now there is discussion among scientific circles about the existence of Moore's law itself. Ultimately, the future of electronic innovations and the works of several people working in material science and physics are in jeopardy. In fact, Moore's prediction of 1965 are technologically unique, who quietly led the silicon revolution with his own law. Thus the future nanotechnology may enhance the present known barriers of Moore's Law.

6.5 Optical Properties of Nanomaterials

Like anyother properties of materials, even the optical properties of nanomaterials are also differ significantly from the those exhibited by their bulk material counterparts. By carefully controlling the size, shape and surface functionality of nanoparticles a wide range of optical properties may be generated with many useful applications.

Surface plasmons (SPs) are coherent delocalized electron oscillations that exist at the interface between any two materials.

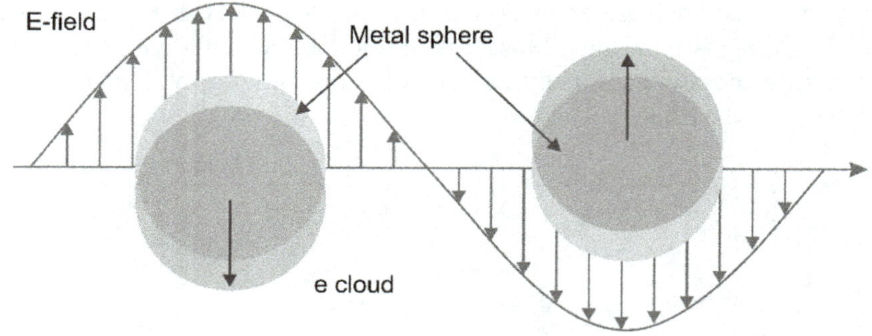

Fig. 6.8 Localized surface plasmon resonance [14]

One of the most interesting properties of nanoparticles in general and metal nanoparticles in particular is their optical properties. Optical properties are totally different from their bulk counterparts and the observed behavior may be attributed to the surface plasmon resonance. In this phenomenon, when light incidents on a metal surface, light wave propagates along the metal surface giving rise to a surface plasmons. In fact, surface plasmons may be considered as conduction electrons that propagate in a direction parallel to the metal/dielectric (or metal/vacuum) interface. When a plasmon is generated in a conventional bulk metal, electrons can move freely in the material and no effect is registered. In nanoparticles, however, the surface plasmon is localized in space and oscillates back and forth in a synchronized manner in a small space and are called as Localized Surface Plasmon Resonance (LSPR). When the frequency of this oscillation is same as that of the incident light, the plasmon is in resonance with the incident light.

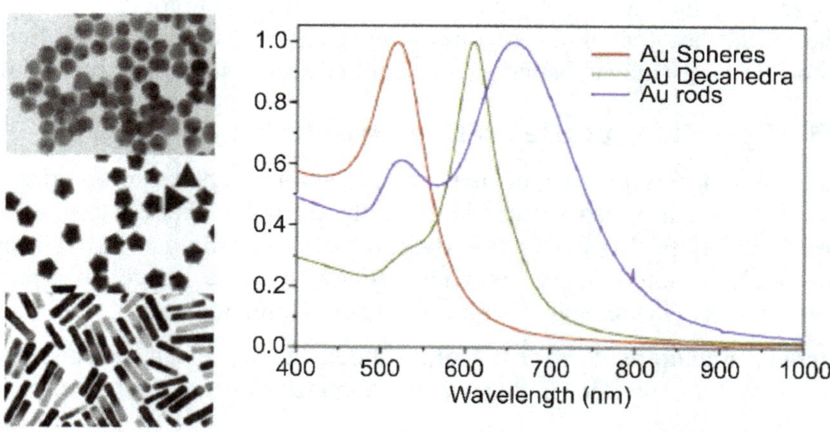

Fig. 6.9 TEM and UV spectra of gold nanoparticle collids [15]

The LSPR effect especially on metal nanoparticles such as gold, silver and aluminum is that they absorb visible light due to the resonance effect of plasmons, As a result, the metal nanoparticles such as gold or silver can display colours which are not found in their bulk form. These colors include red, purple, blue or orange, depending on the nanoparticles' shape, size. Therefore, by tuning the size and shape, the wavelength can be shifted from visible into the infrared region. This property of metal nanoparticles makes them useful for sensor applications.

The surface plasmon resonance (SPR) of the free electrons in the metal nanoparticle is sensitive to the dielectric function of the material and the surroundings in addition to the shape and size of the nanoparticle [16], which can be understood by studying the polarizability.

For a spherical nanoparticle, the polarizability (α) is given by,

$$\alpha = 4\pi\varepsilon_0 r^3 \frac{\varepsilon_1(\omega) - \varepsilon_2}{\varepsilon_1(\omega) + 2\varepsilon_2}$$

where ε_1 is the dielectric constant of the nanoparticle and ε_2 is the dielectric constant of the medium. When the condition,

$$\mathrm{Re}\{\varepsilon_1\} = -2\varepsilon_2$$

is satisfied, the particle exhibits resonance resulting in increase in the absorption.

Therefore, the nanoparticles optical properties are highly dependent on material composition, size, shape and the medium in which the particles are embedded.

6.6 Magnetic Properties

Nanoparticles of magnetic materials have attracted much attention because of their properties often differ considerably from those of bulk materials. Therefore, the magnetic nanomaterials can be used to make materials and devices with new properties. Magnetic nanoparticles have several applications. They include magnetic data-storage devices, magneto resistive read heads, magnetoresistive RAMs, magnetic refrigeration, ferrofluids, in MRI and for targeted drug delivery purposes. In addition magnetic nanoparticles can be used in diagnoses of diseases. Due to their unique chemical and physical properties, incorporating them in biosensors may improve the sensors' sensitivity.

Super Paramagnetism

Magnetic materials exhibit size-dependent magnetic properties ranging from ferromagnetic to paramagnetic and superparamagnetism. The magnetic properties of nanoparticles differ from those of bulk mainly in two aspects. The large surface-to-volume ratio results in a different local environment for the surface atoms leading to the novel magnetic characteristics. The second aspect exhibited by the nanomagnetic materials is the super paramgnetism. Unlike bulk ferromagnetic materials, some of the nanoferromagnetic materials may consist of only a single magnetic domains and these materials may exhibit superparamagnetism. Such type of materials do not exhibit magnetism in the absence of external magnetic field. However, these type of materials quickly become magnetized in an external magnetic field. Further, when returned to a zero magnetic field, these materials quickly revert to the non-magnetic state. The superparamagnetic relaxation has important technological applications. In magnetic data-storage media, it is crucial that the magnetization is stable for several years. In fact, in information storage, the size of the domain determines the limit of storage density [17, 18].

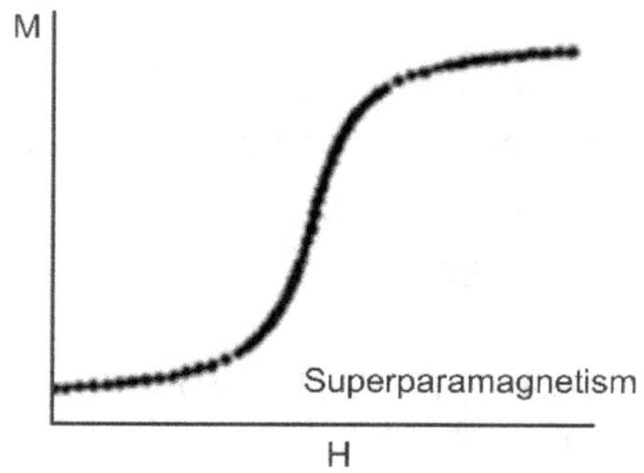

Fig. 6.10 Superparamagnetism

Coercivity

The size of nanomaterial has a great influence on its magnetic behavior. As the size of a particle is reduced, the whole particle becomes a single domain below a critical size. as can be illustrated by considering the coercivity of that material. The size-dependent coercivity of material is shown schematically in Fig.6.11. Firstly, for very small particles with diameters smaller than the critical value of ≤ 1 nm, super paramagnetism is present as shown in Fig. 6.11. Secondly, in the range between 1 nm and critical diameter of a single domain, the moment is stable and the coercivity increases slowly reaching a peak value as shown in Fig.6.11. Finally, as the particle diameter is increased further, slowly the multi-domain region appears, and the coercivity starts decreasing gradually. Therefore, one may understand that the magnetic material exhibits maximum coercivity when the nanoparticle diameter is almost equal to single domain..

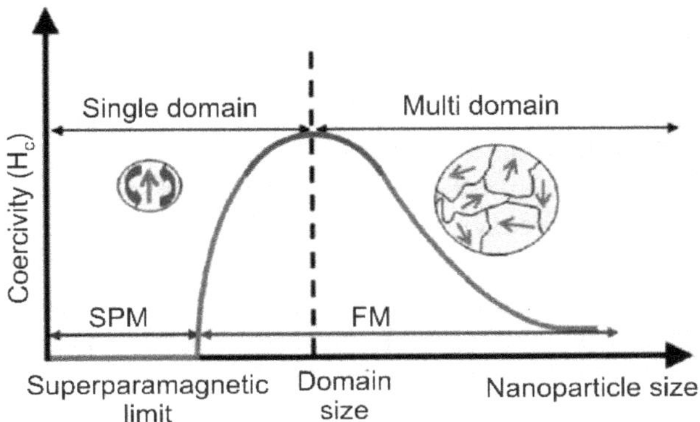

Fig. 6.11 Size-dependent coercivity of magnetic particles [19]

6.7 Thermal Properties

Nanomaterials exhibit interesting physical, chemical, magnetic and thermal properties that are significantly different from the corresponding properties of their bulk materials counter parts. These interesting properties arises because of their high value of surface-area-to-volume ratio. Moreover, the energy associated with the atoms of the nanomaterials is different compared with those of conventional bulk materials. This in turn may lead to the size-dependent interesting thermodynamic properties of nanomaterials.

Thermal conductivity

Heat is transmitted in materials by two different mechanisms:

They process are,

➢ Lattice vibrations (phonons) and

➢ Free electrons.

In metals, thermal transport takes place via the electrons rather than by phonons. In contrast, in the case of nonmetals, phonons are responsible for thermal transport. However, in both metals and nonmetals, especially when they are in nanoscale, quantum confinement and grain boundary scattering effects will influence thermal conductivity values.

In general due to the presence of large number of grain boundaries, phonons scatter thermal energy resulting in exhibiting low thermal conductivity values. Even in the case of nanomaterials also   because of the presence of large number of grain boundaries, they are found to exhibit lower value of thermal conductivity than the coarse grained ones [20]. In nanomaterials, due to reduction in the crystallite size values to almost to nanodimensions, the crystallite size values sometimes become equal to the phonon mean free path values, which are responsible for the transport of thermal energy. In such a scenario, the nanomaterials exhibit totally different thermal properties. The shape of nanomaterials is also having influence on their thermal properties. For example, 1D- nanowires, are found to exhibit ultra low thermal conductivities when compared with Carbon nanotubes. Moreover, in 1D nanowires, quantum confinement of phonons may leadto a strong phonon- phonon interactions. Thus enhanced scattering at the grain   boundaries and phonon – phonon interactions might be responsible exhibiting significantly low thermal conductivity values.

Thermal conductivity $\mathbf{K= (1/3)* vC_v l}$

where v is a particle velocity, l is a free path length, Cv is the heat capacity of unit volume.

In case of bulk materials, $d=l_{phonon}$ or $d \geq l_{phonon}$ (d is diameter of the particle) but in the case of nanomaterials $d < l_{phonon,}$ resulting to cut of phonon spectra and decreasing the thermal conductivity (K) of the material. Therefore in case of nanomaterials, the thermal conductivity is significantly reduced with decreasing size of the particle.

Specific heat

The **specific heat** is the amount of **heat** per unit mass required to raise the temperature by one degree Celsius.

$$C_n = C_b(1-N/2n)^{-1}$$

This is the relationship of the specific heat for nanomaterials and bulk materials at different shapes and sizes.

For example, the specific heat of Ag nanoparticles is plotted against size in the following figure. It can be seen from the figure that the specific heat is found to increase with decreasing size of the nanocrystal. From this, one may conclude that the specific heat capacity varies inversely with the particle size of the nanomaterial. This behavior may be attributed to the fact that the high atomic thermal vibration energies of the surface atoms. The higher fraction of the grain boundaries in nanocrystals also results in higher values of specific heat compared to conventional polycrystals [21].

This is the relationship of the specific heat for nanomaterials and bulk materials at different shapes and sizes.

For example, the specific heat of Ag nanoparticles is plotted against size in the following figure. It can be seen from the figure that the specific heat is found to increase with decreasing size of the nanocrystal. From this, one may conclude that the specific heat capacity varies inversely with the particle size of the nanomaterial. This behavior may be attributed to the fact that the high atomic thermal vibration energies of the surface atoms. The higher fraction of the grain boundaries in nanocrystals also results in higher values of specific heat compared to conventional polycrystals [21].

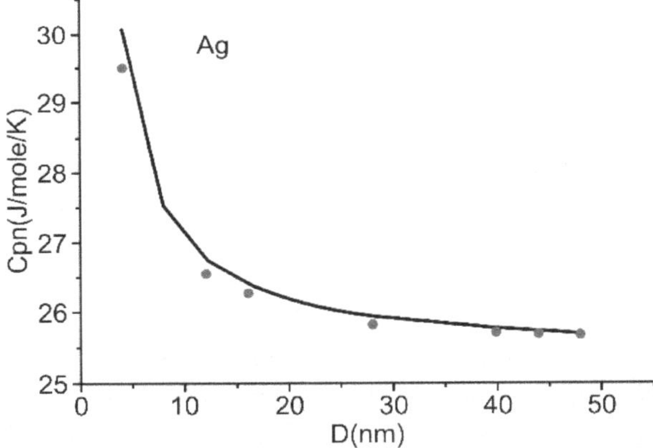

Fig. 6.12 Size dependence of specific heat

6.8   Nanofluids

The conventional liquids such as water and some oils with millimeter or micrometer sized particles were normally using for Heat transfer applications. However, these liquids are having very low thermal conductivity when compared with the solids. Moreover, these liquids cannot work with the emerging miniaturized technologies. Therefore, people started looking for new materials. In this connection new material, nanofluids were discovered.

The stable suspension of nano sized particles in a base fluid is called **nanofluid.**

Typically, nanofluids contain up to 10 vol% of nanoparticles and either oil or water is used as a carrier liquid. The purpose of suspension of **nanoparticles** in various base fluids is to alter the fluid flow and heat transfer characteristics of the base fluids.

➢ non-metallic particles - SiC, TiC, SiO2, TiO2, Al2O3, ZnO, CuO, Fe3O4, SiN and AlN

➢ metallic particles - Cu, Ag, and Au and

➢ different particle shapes such as carbon nanotubes, nanodroplets, nanofibers, and nanorods.

The base fluids commonly used are

➢ Water

➢ Ethylene glycol

➢ Engine oil

➢ Mineral oil

➢ Glycerol

➢ Acetone

Normally, two methods are used for the preparation of nano fluids. They are,

1. Single step process.

2. Two- step processes

1. To reduce the agglomeration of nanoparticles, this single step process is used. In this process, the nanoparticles are preared and dispersed in the base fluid simultaneously. The nanoparticles can be prepared by one of the suitable methods. Therefore, the process of drying, storage, transportation and dispersion of nanoparticles can be avoided. This may help in minimizing the agglomeration process, which in turn improves the nanoparticle dispersion in the base fluid. Inspite of these advantages of the method, there are few disadvantage also. The disadvatages include residues of reactants in the nanofluid due to incomplete reaction. Another problem with the single step process is the low vapor pressure may limit the application of this process.

2. Two step Process: In this method, the nanoparticles are first produced by one of the suitable routes. The second step involves dispersing this nanomaterial into the base fluid with the help of intensive magnetic force agitation. Infact, the two-step process is more economical than the one-step method in producing the nanofluids commercially. The main problem with this method is that, due to the high surface to volume ratio of nanoparticles may result in agglomeration, which in turn may result in decreasing thermal conductivity and increase in clogging of microchannels. Therefore, surfactants are widely used to stabilize nanoparticles in the fluids. In fact, this method is suitable for wide range of particles including oxide particles and carbon nanotubes. Moreover, this method is attractive because of   simple fluid preparation.

Applications of Nanofluids

Nanofluids are used mainly in automobile and the energy industries. Some applications in electrical and electronic industries are mainly for cooling the devices. In mechanical engg. industry, the nanofluids are used for efficient heat transfer and energy generation and process industry energy recovery from flue gases. Nanofluids are extensively used for cooling and heating the buildings. Apart from this nano fluids are also used for thermal storage, solar energy systems, desalination industry, refrigeration, space and defense applications [22].

6.9   Mechanical Properties

The novel properties of nanoparticles are often attributed to the large surface to volume ratio. Surfaces and interfaces can have a strong impact on mechanical properties of nano materials such as hardness and elastic modulus, fracture toughness, scratch resistance, fatigue strength, and hardness.

Strength and Hardness

Strength is the ability of a material to resist deformation caused by external load. The more external loading a material can withstand, the higher strength it will have.

The **Hall–Petch** relation predicts that as the grain size decreases the yield strength of a material increases.

$$\sigma_y = \sigma_0 + Kd^{-1/2}$$

where σ_y is the yield stress , σ_0 is the friction stress, k is the constant and d is the grain diameter.

From this it has been concluded that the strength of materials is the function of grain size. It means that strength of materials is inversely proportional to grain diameter.

Hardness of a material is its ability to resist indentation, penetration or scratching caused by another material. It is always relevant to the surface of a solid, not the entire material.

Subsequently, it was reported also that the **Hall–Petch** relation holds for **Hardness** of the materials and is given by,

$$H = \alpha H_0 + Kd^{-1/2}$$

In the above equation, as H_0 and K are constants, one may write that

$$H \, \alpha d^{-1/2}$$

Therefore, one may conclude that the Hardness of materials is inversely proportional to grain diameter. Thus hardness of nanophase materials increases with decreasing grain diameter.

The Hall-Petch relation predicts the relationship between a metal's yield strength and its average grain size especially when the average grain size 100 nm and larger. It means that, this relationship, may not be applicable for smaller grains, including nano size grains. However, "Inverse Hall-Petch relationship" obeys in the case of metals with very small grain sizes [23]. Further, it was reported that the mechanical properties of a solid depend on its density of dislocations, interface-to-volume ratio and grain size. A decrease in grain size significantly affects the yield strength and hardness. As most of the metals are made up of small crystalline grains, the grain boundaries may slow down the propagation of defects when the material is stressed. If these grains are nanoscale size, the interface area (grain boundary) within the material greatly enhancing its strength.

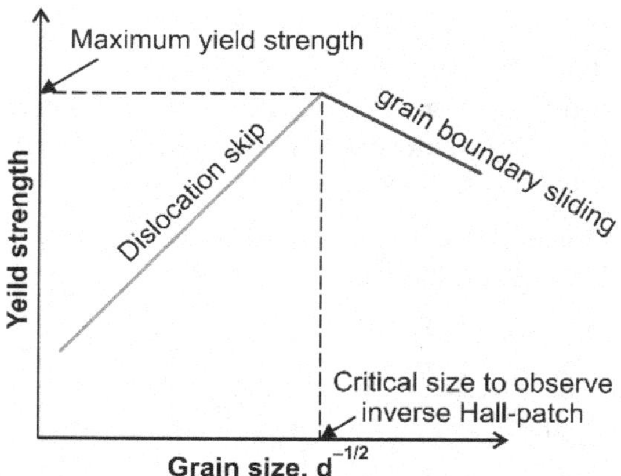

Fig. 6.13 Schematic representation showing inverse Hall-Petch effect below a critical grain size

Superplasticity

According to Hooke's Law, the strain of the material is proportional to the applied stress within the elastic limit of that material.

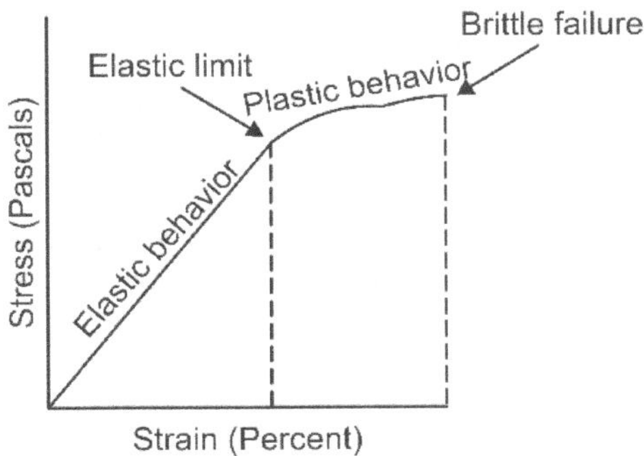

Fig. 6.14 Stress-Strain curve

Some metals and alloys deform uniformly well beyond their usual breaking point, usually over about 600%   is called **superplasticity.**

Examples: Some fine-grained metals and ceramics (grain size should be less than 20μm.)

The grain boundary structure, boundary angle, boundary sliding and the movement of dislocations are important factors that determine the mechanical properties of the nanostructured materials. One of the most important applications of nanostructured materials is in superplasticity, the capacity of a polycrystalline material to undergo extensive tensile deformation without fracture. Grain boundary diffusion and sliding are the two key requirements for superplasticity [24]. The enhanced inter diffusion among the grains in nanocrystals helps the grain boundaries slide by one another, thereby improving the superplasticity.

References

1. G. Cao, Nanostructures and Nanomaterials: Synthesis, Properties & Applications, Imperial College Press, 2004.

2. H. Ogawa, M. Nishikawa, A. Abe, Journal of Applied Physics 53 (1982) 4448.

3. H. Luth, Surfaces and Interfaces of Solid Materials, Springer, Heidelberg, 1995.

4. P. Knauth, J. Schoonman, Nanostructured Materials: Synthesis Methods, Properties and Applications, Springer 2002.

5. M. Zhang, M. Yu. Effremov, F. Shiettekkatte, E.A. Olson, A.T. Kwan, S.L. Lai, T. Wisleder, J.E. Greene, L.H. Allen, Phys. Review B 62 No.15 (2000) 10548.

6. K.J. Hanszen, Z. Phys. 157 (1960) 523

7. Glenn Clayton, Nanoscience and Nanotechnology, ED-Tech Press, 2018

8. J. Rupp, R. Birringer, Phys. Rev. B, 36 (1987) 7888

9. 1. Liliana Marinescu et al, "Optimized Synthesis Approaches of Metal Nanoparticles with Antimicrobial Applications", Journal of Nanomaterials, 2020 .

10. Jiao Sun, Fan Wang, Yue Sui, Zhennan She, Wenjun Zhai,Chunling Wang, and Yihui Deng, Effect of particle size on solubility, dissolution rate, and oral bioavailability: evaluation using coenzyme Q_{10} as naked nanocrystals, Int J Nanomedicine. 2012; 7: 5733–5744.

11. Guozhong Cao, Nanostructures & Nanomaterials: Synthesis, Properties & Applications, Imperial College Press, 2004

12. Madhuri Sharon, History of Nanotechnology: From Prehistoric to Modern Times, Scrivener Publications, 2019

13. Jerry Wu, Yin-Lin Shen, Kitt Reinhardt, Harold Szu and Boqun Do, A Nanotechnology Enhancement to Moore's Law, Applied Computational Intelligence and Soft Computing, 2013

14. Kelly, K.L., et al., The optical properties of metal nanoparticles: The influence of size, shape, and dielectric environment. Journal of Physical Chemistry B, 107(3): p. 668-677, 2003.

15. Borja Sepúlveda et al., "LSPR-based Nanobiosensors", Nano Today, Vol. 4 (3), 244-251, 2009

16. G. Mie, A contribution to the optics of turbid media, especially colloidal metallic suspensions, Ann Phys, 25 (1908), pp. 377-445.

17. R.G. L. Audran, A.P. Huguenard, U. S. Patent, 4302523 (1981)

18. R.F. Ziolo, U.S. Patent, 4474866 (1984)

19. Mehrmohammadi, Mohammad, Yoon, Ki youl , Qu, Min, Johnston, Keith and Emelianov, Enhanced pulsed magneto-motive ultrasound imaging using superparamagnetic nanoclusters. Nanotechnology. Vol.22, 2011.

20. Wu, H., Carrete, J., Zhang, Z. et al. Strong enhancement of phonon scattering through nanoscale grains in lead sulfide thermoelectrics. NPG Asia Mater 6, e108 (2014).

21. Ph. Buffat, J. P. Borel, Phys. Rev. A, 13 (1976) 2287

22. Mohamoud Jama et al., Critical Review on Nanofluids: Preparation, Characterization, and Applications, Journal of Nanomaterials, 2016.

23. Tschopp, M.A., Murdoch, H.A., Kecskes, L.J. et al. "Bulk" Nanocrystalline Metals: Review of the Current State of the Art and Future Opportunities for Copper and Copper Alloys. JOM 66, 1000–1019 (2014).

24. J. Karch, R. Birringer, H. Gleiter, Nature, 330 (1987) 556.

Chapter 7

Applications of Nanomaterials

Nanotechnology is a field of applied science concerned with the control of matter at dimensions of roughly 1 to 100 nanometers (nm). Materials at the nanoscale have different properties compared to the same chemical substance in a larger size. These unique properties at the nanoscale make nanomaterials to have several applications in various fields such electronics, computers, biotechnology, textiles, catalysis, paintings and coatings, personal care products, energy savings, alternative energy supplies, efficient use of raw materials, environmental protection, agriculture applications and medical breakthroughs. This chapter will provide a brief description of the applications of nanotechnology in a wide spectrum of fields.

7.1 Nanoelectronics

Nanotechnology plays vital role in the field of electronics. The resulting field developed from the combination of nanotechnology and electronics is called as "NANOELECTRONICS". It is also defined as the science which deals with the electronic components manufactured and engineered at a molecular scale. The main target of this technology is to develop electronic products and circuits at super miniature level. It also provides faster, smaller and more portable systems. Some of the Consumer Products which use Nanotechnology are computer hardware, display devices, mobile & communication products, audio products, digital cameras etc., The continuously evolving applications of nanotechnology include:

Nanotransistors

The most widely used products in the Electronics industries is a transistor. It is the major component and basic building block of any electronics equipment. As transistor technology advanced, their dimensions were reduced from millimeter to micrometers (μm) and from micrometers to the nanometers (nm) scale, so that more and more of them could be included in most modern and new electronic systems. Today, billions of transistors are used in our smart phones, tablet and personal computers, supercomputers and other electronic systems that have shaped the world we live in [1]. Smaller, faster and better transistors may mean that soon your computer's entire memory may be possible to store everything on a single tiny chip.

Memory Storage

With nanotechnology, magnetic random access memory (MRAM) are possible. This may help in booting the computers almost instantly. In fact, magnetic tunnel junctions can quickly and effectively save data during a system shutdown.

Ultra-High Definition Displays

The production of displays with low energy consumption might be accomplished using carbon nanotubes (CNT). This is possible because carbon nanotubes are electrically conductive and due to their small size (nanometers), they can be used as field emission displays (FED). The working principle of these devices is similar   to that of the cathode ray tube, but on a much smaller length scale [2].   Therefore, the ultra-high definition displays with more energy efficiency are possible with carbon nanotubes.

Fig. 7.1 Ultra-high definition displays (QLED TV)

Nanosensors

Nanosensors are extremely small sensing devices that are capable of detecting and responding to physical stimuli with dimensions in the range of nanometer. Nanosensors are biological, chemical, or surgical sensory points used to convey information about nanoparticles to the macroscopic world. Nanosensors have applications in defense, medical, healthcare and consumer products. The nanosensors are also used to detect airborne chemicals, viruses, bacteria, measuring the temperature of living cells and measuring temperature of nanofluids etc. Apart from this, nanosensors can also respond in a faster way, can have better signal-to-noise, can have more accurate data, increased data density, less impact on the phenomenon being measured.

Nanorobotics

A nanorobot is a tiny machine designed to perform a specific task or tasks repeatedly and with precision at nanoscale dimensions. The terms nanobot, nanoid, nanite, nanomachine, or nanomite have also been used to describe such devices currently under research and development. Nanorobotics has a wide application in Medical field. Nanorobotic is especially used for the removal of kidney stone, dental treatment, elimination of defected part in the DNA structure, diagnosis and drug delivery in cancer treatment etc. The Nanorobotic future is excellent and bright. The advantage of this technology is that it reduces the risk and minimizes the cost of surgery and makes the life of a surgeon easy.

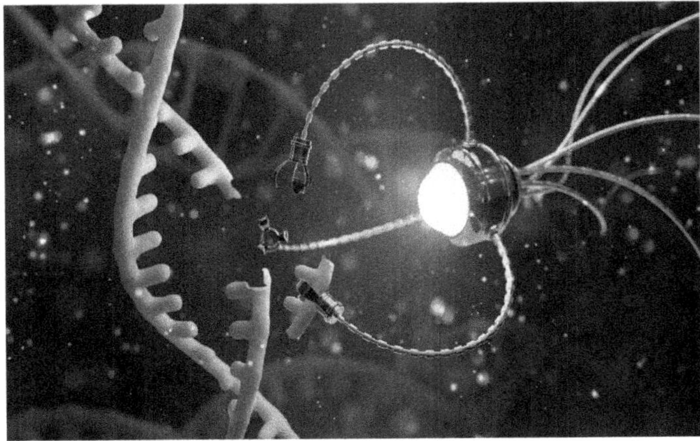

Fig. 7.2 Nanorobotics (Image Credits: K_E_N/shutterstock.com)

Flexible electronics

Nanowire electrodes because of their size help in designing flat panel displays. These are very much thinner than the currently available flat panel displays. The transistors, made with nanowires are normally used for making display devices. These are assembled on glass or thin films of flexible plastic. E-paper, displays on sunglasses and map on car windshields [3].

The electronic devices that are flexible, bendable, foldable, roll able and stretchable have entered into every walk of life. In fact these devices have been integrated into a variety of products, such as medical, aerospace applications along with Internet of Things. Semiconductor nano-membranes are normally used for the manufacture of Flexible electronic devices and are extensively used in smart phone and e-reader displays. Other materials such as graphene and cellulosic nanomaterials are also used for the manufacture of flexible electronic devices. Using, flat, flexible with lightweight and non-brittle characteristics makes these novel materials useful in variety of smart products. Other applications include Flash memory chips for smart phones and thumb drives; ultra-responsive hearing aids; antimicrobial & antibacterial coatings on keyboards and cell phone casings. The conductive inks for printed electronics for RFID & smart cards & smart packaging, flexible displays for e-book readers are the other applications of these novel electronic devices [4].

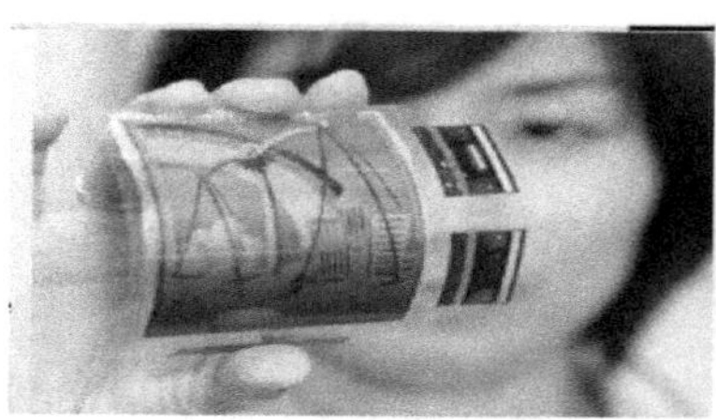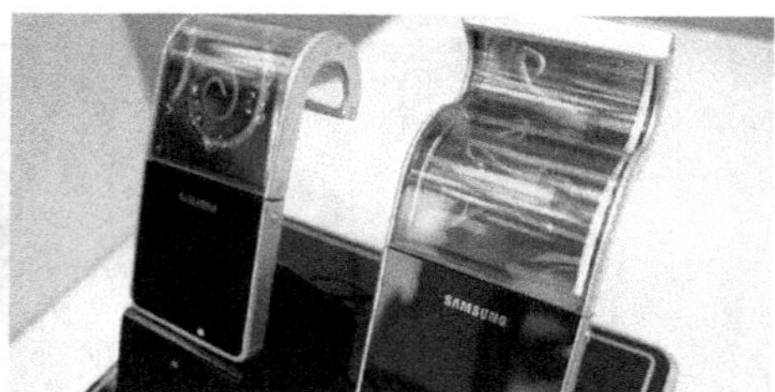

Fig. 7.3 Flexible displays

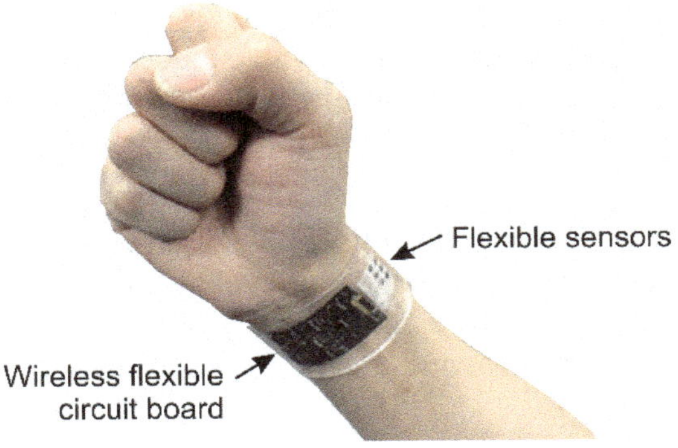

Fig. 7.4 Flexible Sensors [5]

7.2 Nano Textiles

Textile manufacturers have begun to use nanomaterials in their products. The unique properties of nanoparticles and nanofibres have been used in designing fabrics with excellent mechanical strength, chemical resistance, water repellence, antibacterial properties and a wealth of other properties which are unattainable by any other means. A first generation of nano - textiles benefitted from *nano finishing*: Coating the surface of textiles with nanoparticles is an approach used for the production of highly active surface. Some of the applications of the nano textiles are UV-blocking, antimicrobial, antistatic, flame retardant etc., Water and oil repellency, wrinkle resistant and self-cleaning [6] are some of the properties influenced by the nanomaterial applications.

Fig. 7.5 The superhydrophobic properties of the lotus leaf are responsible for water- and   stain-repellent clothing

Water droplets skid off both lotus leaves and nanotextiles. Water droplets just skid off lotus leaves because they have super hydrophobic surface as the surface of the leaves are coated with hydrophobic wax nanocrystals which repel water droplets and minute dust particles. Mimicking this, textile scientists have prepared stain and wrinkle proof textile material by coating the fabric with thin layer of extremely hydrophobic silicone nanofilaments. The unique spiky structure of the filaments creates a 100% water proof coating. Water droplets rest on the silicone coating and slide off from the cloth when the garment is just tilted.

Porous Polytetrafluoroethylene (PTFE) is used to produce clothing that protects the wearer from rain, wind and even snow. PTFE treated fabrics act like nanofilters – the protective inner layer protects the fabric from stains from the body's sweat and oil and the outer layer protects the fabric from damage.

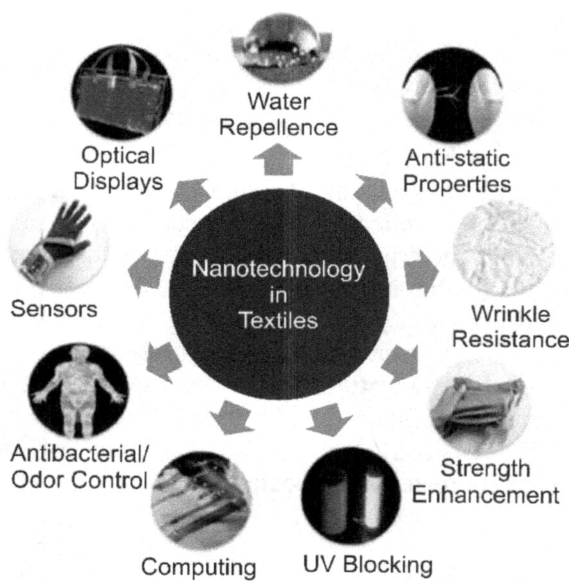

Fig. 7.6 Applications of nanotechnology in textiles [6]

Anti-bacterial and anti-fungal silver nanoparticles are used to produce a range of clothing impregnated with silver nanoparticles. Socks made of such material are useful in inhibiting the growth of bacteria and fungi which cause toe nail infections, foot odor and painful cracks in the heel. The antimicrobial properties are exerted by nano-silver, while nano-metal oxide coatings are responsible for blocking the dangerous ultra violet rays. The metal oxide coatings are also responsible for self-cleaning and flame-retardant properties. Further, UV-protection ability in textiles and sunscreens, and antibacterial finishes in medical textiles and inner wears are developed using Zinc oxide nanoparticles embedded in polymer matrices like soluble starch.

How Nanomaterials are used in Textiles ?

The most common way of using nanomaterials is by making composite with conventional fibre with nanoparticles to enhance performance of ordinary clothes. In fact, nano textiles may be responsible for some of the novel properties to the fabric. Some examples are given in the table 7.1.

Table 7.1 Properties of nanomaterials used in textiles [7]

Nanomaterial	Properties
Carbon black nanoparticles or nanofibers	Abrasion resistance
	Higher tensile strength
	Good chemical resistance
Carbon nanotubes	Exceptionally strong (100x tensile strength of steel)
	Lightweight
	Electrically conducting
	Thermally conducting
Metal oxide nanoparticles	Photocatalytic
	Electrically conductive
	Antimicrobial
	Solar cells
Clay nanoparticles	Electrical resistance
	Chemical resistance
	Fire retardant
	UV shielding

Nano-Textile Products

- Fabrics for sports people with improved mechanical properties and odour-reducing antibacterial properties are the important ones.
- Special clothing for medical professional including for antimicrobial wound dressings.
- Personal Protective Equipment (PPE) Kits especially for chemical and heat protection.
- Special textiles for Military personnel including body armor, radio shielding and camouflage.
- Special textiles for people working in the electronic industry made up of nanofibres.

7.3　Nanotechnology in Automobile Industry

The main advantages of applying nanotechnology in automobiles include providing lighter and stronger body parts (to enhance safety and fuel efficiency), improving fuel consumption efficiency, and therefore achieving a better performance over a longer period. For this purpose, high strength- to- weight ratio materials. In this connection, Nanotechnology plays a major role in realizing many of these objectives especially in enhancing the vehicle performance, convenience and safety [8]. Some of the applications of nanotechnology in the automobile industry are given below:

Applications

The basic trends that nanotechnology enables for the automobile sector are,

a) Lighter but stronger materials for better fuel consumption and increased safety.

b) Nanoscale additives in polymer composite materials are being used in bicycles, motorcycle helmets, automobile parts, luggage, and power tool housings, making them lightweight, stiff, durable and resilient.

Carbon nanotube based composites are being examined as a replacement for the automobile frames due to their high strength and reduced weight. The use of materials with high strength-to-weight ratio for body frames will essentially make the automobile crash resistant as well as decrease the fuel consumption due to reduced weight.

➤ improved engine efficiency and fuel consumption for gasoline-powered cars (catalysts; fuel additives; lubricants and better batteries, Fuel Cells)

➤ reduced environmental impact from hydrogen and fuel cell-powered cars

➤ improved and miniaturized electronic systems

➤ better economies (longer service life; lower component failure rate; smart materials for self-repair)

➤ Lotus Coatings

The lotus effect refers to the very high water repellence exhibited by the leaves of the lotus flower. Some nanotechnologists have developed treatment, coatings, paints and other surfaces that can stay dry and clean themselves in the same way as lotus leafs.

Clear nanoscale films on mirrors, side panel glasses (windows), head lights and other surfaces can make them water- and residue-repellent, antireflective, self-cleaning, resistant to ultraviolet or infrared light, antifog, antimicrobial, scratch-resistant, or electrically conductive.

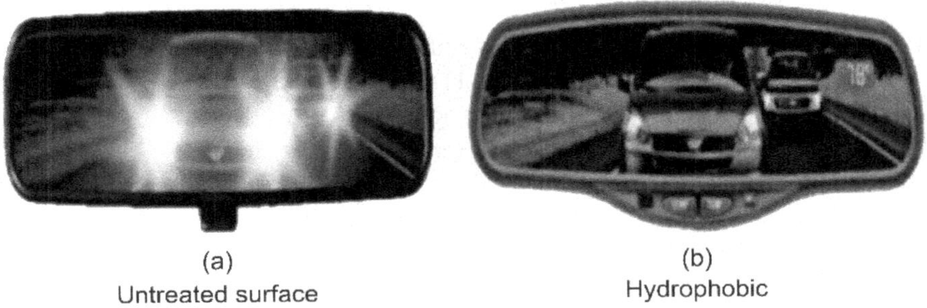

(a)
Untreated surface   (b)
Hydrophobic

Fig. 7.7 Effect of electrochromic coating demonstrated by the example of a rear view mirror

➢ Nanofilters for clean air in the interior of the car:

The automotive industry tends to especially provide to its customers enhanced comfort besides improved safety and reduced fuel consumption. High-quality interior air filters (Novel filters covered with nanofibres) remove a great amount of particles (e.g. pollens, spores, industrial dust) as well as odours from air entering the car.

With nanotechnology future trends in automobile sector

Nanotechnology offers developing a new multifunctional materials that will help in building and maintaining lighter, safer, smarter and more efficient vehicles, aircraft, spacecraft and ships. The overall benefits of transport sector may be outlined here.

➢ Polymer nanocomposites may be used for developing high-power rechargeable battery systems; thermoelectric materials for temperature control; lower rolling-resistance tires; high-efficiency/low-cost sensors and automobile electronics.

➢ Development of Nanoscale sensors, communications devices for the enhanced transportation infrastructure that can communicate with vehicle-based systems so that the drivers can maintain lane position, suitable changes in the travel routes to avoid congestion and development onboard electronics so as to improve the interface with the driver.

➢ Use of nanotechnology for the development of lightweight, high-strength materials for spacecrafts mainly for saving expensive energy needed for launching a spacecraft into orbit.

➤ Nanotechnology-enabled lubricants and engine oils also significantly reduce wear and tear, which can extend the lifetimes of spare parts and others used in automobiles[9] .

7.4 Nanotechnology for Water Treatment

Water is the most important asset of human civilization, and potable water supply is a basic human necessity. Demand escalates due to population growth, global climate change and water-quality deterioration. Although large number of seas and oceans are there only 2.5% of the world's requirement harness fresh water. Moreover, 70% of fresh water is frozen in the form of ice. In fact, only <1% of fresh water can be used for drinking. Globally, >700 million people do not have access to potable water. Therefore, water treatment plays an important role in the lives of mankind.

Importance of Nanotechnology in Water Purification

➤ Nanotechnology provided innovative solutions for water treatment. Nanotechnology-enabled processes are highly efficient, modular, and multifunctional in nature, and they provide high performance, affordable water and, wastewater treatment solutions.

➤ Nano materials due to their high value of surface to volume ratio controls the pollutants and bacteria.

➤ Nanotechnology can be used for the purification of water sources in an economic way.

➤ Waste water treatment is the biggest problem for the people working in water works department. However, methods using Nanotechnologies are more advantageous in treating waste water leading to reduction in labor, time, and expenditure to industry solving various environmental issues.

Some nanotechnology applications for water and wastewater treatment are discussed below

a) **Nanoadsorption:** Adsorption is a surface process wherein pollutants are adsorbed on a solid surface. However, the nanoadsorbents, because of their small size, catalytic potential, high reactivity, large surface area, ease of separation make them ideal adsorbent materials especially for the treatment of wastewater

b) **Nanostructured catalytic membranes (NCM):** Nanostructured catalytic membranes are widely used for water contamination treatment. Nanostructured membranes are already widely applied to remove dissolved salts and micro-pollutants, soften water and treatment of wastewater. Some of the functions of nanostructured membranes include, decomposition of Organic pollutants, make the microorganisms inactive, and physical separation of water contaminants. The N-doped ZnO nanostructured material is very efficient in removing water contaminants by using visible light irradiation.

c) **Membrane Filtration Technology:** Nanofiltration is used to remove pesticides and other organic contaminants from surface and ground waters to help insure the safety of public drinking water supplies. It is also capable of removing bacteria and viruses as well as organic-related color without generating undesirable chlorinated hydrocarbons.

Therefore, nanomaterials are having a great potential in waste water treatment. In fact, high surface area to volume ratio is used efficiently in removing toxic metal ions. The other application of the nanomaterials include disease causing microbes, organic and inorganic solutes from water. Nanotechnology based techniques are very much useful, cost effective, efficient, durable and eco-friendly. These methods are less time taking and low energy consuming techniques when compared with the conventional bulk materials based methods [10].

7.5 Nanotechnology for Environment Protection

Nanotechnology is an emerging field that covers a wide range of technologies. It plays a major role in the development of innovative methods to produce new products and chemicals with improved performance resulting in less consumption of energy and materials and reduced harm to the environment as well as environmental remediation [11]. Nanotechnology is used for cleaning the environment. This includes examining the possibility of improving existing pollution, control methods, improving the existing manufacturing methods to reduce the new pollution, and making alternative energy sources are more cost effective.

Some of the potential applications of nanotechnology are,

➢ Early application of nanotechnology is remediation using nanoscale iron particles. Zero-valent iron nanoparticles disperse throughout the body of water to remediate soil and water contaminated with chlorinated compounds and heavy metals. This method can be more effective and cost significantly less than treatment methods that require the water to be pumped out of the ground.

➢ One of the environmental applications of nanotechnology is in the water sector. Normally, Carbon nanotube membranes are used for reducing the desalination costs. Similarly, nanofilters are also used for cleaning the ground water contaminated with chemicals and hazardous substances [12].

➢ For energy-efficient desalination, Molybdenum disulphide (MoS_2) membranes filters are used. These are found to filter two to five times more efficiently, than those of conventional filters.

➢ For cleaning industrial water pollutants in ground water nanoparticles are used. This method is most cost effective.

➢ A paper towel made up of nanofabric can absorb 20 times higher than ordinary clothes.   A magnetic water-repellent nanoparticles is used for cleaning oil spills and is far better than the mechanically removal method.

➢ Silver and titanium dioxide with antimicrobial characteristics may be helpful in removing pathogens so that waterborne diseases can be prevented.

➢ Nanofilters could be employed in automobile tailpipes and factory smokestacks so that   contaminants can be prevented entering the atmosphere. This also helps in decreasing the greenhouse gases.

➢ Nanosensors could also be developed to detect toxic gases at very low concentrations in the atmosphere.

➢ For many household products such as degreasers and stain removers; environmental sensors, air purifiers, and filters; antibacterial cleansers, etc., nano- engineered materials make superior.

➢ It has been found that an array of silicon nanowires embedded in a polymer results in low-cost but high-efficiency solar cells. This may help in reducing the cost for the production of electricity.

➢ High efficiency solar cells, hydrogen storage for fuel cell, etc., are being developed as environment friendly fuel resources to drive future automotives.

7.6 Nanotechnology for Energy Applications

The majority of the world's energy comes from fossil fuels – primarily from coal, oil, and natural gas. Today, we are using fossil fuels faster than we are finding them. Energy sources widely available are dependent on fossil fuels which are not sufficient to fulfill the need of growing world. Infact, the Oil Depletion Analysis center (ODAC) predicts that in the near future the demand for fossil fuels will far exceed the Earth's supply[13].

Nanotechnology to fulfill the demand

Most popular and widely available inexhaustible and cleanest of all the renewable sources of energy is solar energy. Solar power as alternative energy resources can be used through Photovoltaic technology. Photovoltaic technology has ability to create high-efficiency solar cells which is going to become a key strategy to meeting growing world energy needs. High-efficiency organic photovoltaic (OPVs) is developed currently using nanotechnology to meet this challenge [14].

Solar Power is abundant resource for renewable efficient energy

The conventional solar cells have low efficiency. New solar cells with nanomaterials have a higher efficiency than Silicon based presently existing solar panels. Further development in the solar power is the usage of latest nanomaterials such as quantum dots, nanowires, nanofilms, nanotubes etc.,. These may help in enhancing the efficiency of the solar cells. Nanotechnology can also offer new applications such as flexible panels. Researchers at the University of Toronto are using nanotechnology to develop solar panels capable of harnessing not only the visible light from the sun, but also the infrared spectrum as well, thus doubling the energy output.

Photovoltaic and photosensitive are the alternatives for energy given by nanotechnology is environment friendly and clean energy as nanomaterial used are of low friction Nano lubrications, lightweight Nano composites and energy efficient nanomaterial for thermal insulation. Some of the advantages using nanotechnology are,

➤ An improved efficiency of lighting and heating

➤ Better electrical storage capacity

➤ A reduction in the amount of pollution from the use of energy

Nanotechnology for Energy- Thermoelectricity

Thermoelectricity is the conversion of heat to electricity and vice versa. Thermoelectric devices have been used for almost a century, however their efficiencies are very poor. Certain Nanomaterials have been developed which have efficiencies three or four times greater than the previous best materials. The reason is in the nanostructured materials the heat transfer is very slow, while still allowing electrons to move freely. By combining several of these thermocouples of nanostructured materials, thermopiles capable of generating reasonable quantities of electricity [15] is obtained.

In the future it is envisaged that such thermopiles could convert waste heat from vehicle engines and exhaust to Electricity, removing the need for an alternator to power electrical components and recharge their battery. Thermoelectric devices have other useful applications apart from generating electricity. In fact, the first commercial application of nanostructured thermoelectric materials is in the computing industry for cooling microprocessors.

Nanotechnology for energy-Hydrogen

Energy needs of mankind are met by combustion of coal, gas and oil. However, this is not the efficient way to extract energy from fuels. Fuel cells to some extent may be useful in generating electrical energy. Nanotechnology can offer solutions to material costs, fuel cell efficiency and storage of hydrogen feedstock. Hydrogen has the potential to provide energy without burdening the environment. Hydrogen is also the favorite to replace liquid fossil fuels for powering vehicles.

7.7   Nanotechnology in Medicine

Nanotechnology is having abnormal applications in medical field including diagnostics and therapeutics. Nano technological devices not only helps in diagnosis of cancer and infectious diseases but also in early detection of the disease. Latest developments in nanotechnology are also beneficial in therapeutic field such as drug delivery and protein delivery. Some of the applications of nanotechnology in medicine as diagnostics and therapeutics are explained below.

Nanotechnology in health and medicine

Application of nanotechnology to the fields of health and medicine is discussed in this topic. The medical applications of nano science are innumerable and the projected benefits to the mankind are unimaginable. In fact using nano medicine, it is possible to detect and prevent the diseases. Apart from this entire diagnosis process of a disease can be improved tremendously.

Drug Delivery

In nanotechnology nano particles are used for site specific drug delivery. In this technique the required drug dose is used and side-effects are lowered significantly. This highly selective approach can reduce costs and pain to the patients. Thus variety of nano particles such as dendrimers, nano porous materials etc., find applications. For example, micelles obtained from block co-polymers, are used for drug encapsulation. Similarly, nano electromechanical systems are utilized for the active release of drugs. Iron nano particles or gold shells are finding important application in the cancer treatment. Tissue damage by drugs can be prevented with drug delivery, by regulated drug release. With drug delivery systems larger clearance of drug from body can be reduced by altering the pharma kinetics of the drug.

Nano medicine for Cancer treatment

Due to the small size of nano particles, they can be of great use in oncology, particularly in imaging. For example, quantum dots, can be used in conjunction with magnetic resonance imaging, to produce exceptional images of tumor sites.

Another drug delivery system that may one day replace chemotherapy is **the Kanzius RF therapy.** In this therapy, gold nanoparticles are attached to cancer cells and are subjected to radio waves. As the metal absorbs the energy from the radio waves more efficiently than living tissue, the gold nanoparticles get heated faster and more efficiently by dielectric heating and the cancer cells are 'cooked' inside the body and killed while the healthy cells are left unaffected.

Thus new nano engineered materials that are being developed for effective in diagnosing, preventing and treating cancer.

Nanotechnology in Protein & Peptide Delivery

Protein and peptides are known as biopharmaceuticals and using their behavior with various parts of the body and is responsible for various types of diseases. In fact, targeted delivery of these materials is a new engineering field and these materials are known as dendrimers.

Ophthalmological Applications of Nanomedicine

Some other applications of nanotechnology include   treatment of oxidative stress; measurement of intraocular pressure; use of nano particles for the treatment of prevention of scars after glaucoma surgery, and for treatment of retinal degenerative disease using gene therapy, regenerative nano medicine etc.,. A novel nanoscale-dispersed eye ointment (NDEO) for the treatment of severe evaporative dry eye has been successfully developed [17]. This technique is proved to be very much useful in restoring the normal corneal and conjunctival morphology and is safe for ophthalmic applications.

Nanotechnology in the treatment of neurodegenerative disorders

One of the important applications of nanotechnology is the treatment of neuro degenerative disorders. For this purpose, the functioning of various types nano carriers including dendrimers, nano gels, nano emulsions, liposomes, polymeric nano particles, solid lipid nano particles, and nano suspensions have been studied.

For the delivery of Central Nervous System (CNS) therapeutics, Nanotechnology plays an important role especially in developing new diagnostic and therapeutic tools. Nanotechnology may also help in limiting and even and reversing neurological states. Nanotechnology may also help in the regeneration of damaged neurons. It is also possible to provide neurological protection and may also facilitate the delivery of drugs across damaged portion of the blood–brain portion.

Tissue Engineering

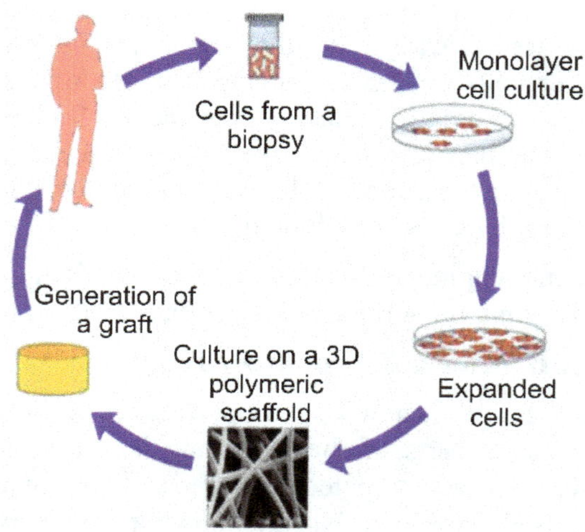

Fig. 7. 8 Basics of tissue engineering [18]

Tissue engineering is a biomedical engineering discipline that uses a combination of cells, engineering, materials methods and suitable biochemical methods to restore, maintain, improve, Thus, in tissue engineering, nanotechnology can be used to reproduce or repair damaged tissues. For example, by using suitable nanomaterial-based scaffolds and growth factors, artificially stimulated cell proliferation, in organ transplants or artificial implants therapy nano technology can be useful. Carbon nanotubes (CNTs) can be used as scaffold material to heal broken bones as they have high mechanical strength and flexibility.

In view of this, one may conclude that nanotechnology offers important tool with a great impact on many areas in medical field. Further, the tissue engineering also provides an opportunity for not only to improve the medical field but also creates new devices and technologies. Many scientific as well as economic activities are expected to accelerate medical research and development. Several medical devices have already developed using nanotechnology.

7.8 Nanotechnology in Cosmetics

The application of Nanotechnology in cosmetic formulations is considered as the very interesting and important field and requires a lot work. In general, manufacturers of Cosmetics normally focus on harmful sunlight and UV protection of skin, deeper skin penetration, increased color, and many more. The applications of nanotechnology can be found in various types cosmetic products including moisturisers, hair care products, make up and sunscreen etc., Some of the recent developments are,

➢ Titanium dioxide and zinc oxide nanoparticles are normally used to block ultraviolet rays. However, we are aware of nanoparticles generated by IVY plants are found to be more effective in blocking ultraviolet rays.

➢ Proteins derived from stem cells are normally used as anti- aging creams. These proteins are encapsulated in nanoparticles liposomes and are delivered to the part of the skin where it is to be applied.

➢ Skin care lotions, wherein the nutrients nanoparticle material in the form of a liquid is to be encapsulated and painted. The small size of the nanoparticles, compared with particles in conventional emulsions, allows the nanoparticles to penetrate deeper into the skin, delivering the nutrients to more layers of skin cells.

➢ Nanopowders are extremely small in size and therefore can be extensively used in the cosmetic industry. Because of smaller size absorption becomes better and therefore less drug is needed for the treatment. As less material is used side-effects are almost negligible.

➢ Nanoscale lipstick last longer, smear less, reflect light and glow in vibrant colors. Therefore nanoscale lipsticks are more advantageous.

➢ The use of nanomaterial implants is beneficial because their anti-bacterial and anti-odour functionality.

➢ Nano-gold, nano-silver particles have anti-bacterial and anti-fungal characteristics. Hence they are used in many personal care products.

➢ Silver nanoparticles dispersed in soaps impart the double advantage of killing germs and increasing effectiveness in removing dirt particles from the skin.

➢ Fullerine based facial creams are being developed to effectively using the anti-oxidation property of C_{60}. The purpose of these anti-oxidants is to protect the hair and skin from ageing.

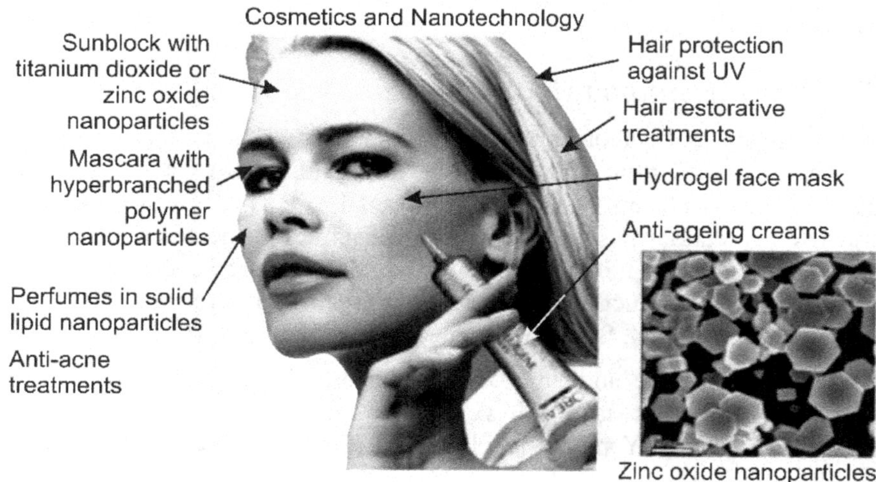

Fig. 7.9 Applications of Nanotechnology in cosmetics and a picture of zinc oxide nanoparticles in cosmetics [20]

7.9 Nanotechnology in Agriculture

Nanotechnology in agriculture has gained good momentum during the last decade. Nanotechnology provides new agrochemical agents and new delivery mechanisms to improve crop productivity and also helps in reducing the usage of pesticide. Nanotechnology can also increase agricultural production.

Some of the applications of nanotechnology in agriculture sector are [21]:

➢ Nanoformulations of agrochemicals are developed for pesticides and fertilizers purposes.;

➢ Nanosensors are used in crop protection in the form of identification of diseases.

➢ Nanodevices are normally used for the genetic engineering of plants;

➢ Nanotechnology is also used for animal health and poultry production.

➢ Post-harvest management is a big problem and as such nanotechnology is extensively is used.

Apart from this, nanotechnology also plays an important role in increasing crop production, food processing and packaging. It is possible to enhance the agricultural productivity through nanomaterial-induced genetically improved plants, gene delivery of molecules at cellular/molecular levels in animals and plants. Nanoparticle-mediated

DNA transfer in plants for the development of insect-resistant plants. Using nanotechnology, it is possible for the precise delivery of pesticides for plant growth.

For the drug and nutrient delivery to fisheries and other livestock, nanoparticles, nanobrushes, and nanomembranes are extensively used. Nanotechnology is successfully used in maintaining freshness, quality, and shelf life of stored products.

The applications of nanomaterials in agriculture sector aimed at time management of various activities so that overall production increases. In view of this, nanotechnology has many applications in different stages of crop related matters. They include production, processing, storing, packaging and transport. Nanotechnology revolutionizes the both the agriculture and food industries. The use of nanomaterials is quite new in agriculture and it requires additional research. Before commercialization and field application, toxicity of nanomaterials is to be evaluated [22].

7.10 Nanotechnology in the Food Industry

Introduction

Over the past few years, nanotechnology has increasingly been considered as to be attractive technology that has revolutionized the food sector. It is observed that these materials have unique properties such as high surface to volume ratio and other properties like color, solubility, strength, diffusivity, toxicity, magnetic, optical, thermodynamic, etc [23, 24]. The rising consumer concerns about food quality and health benefits are impelling the researchers to find the way that can enhance food quality while disturbing least the nutritional value of the product.

Nanotechnology offers complete food solutions from food manufacturing, processing to packaging. Nanomaterials bring about a great difference not only in the food quality and safety but also in health benefits that food delivers. Food nanostructured ingredients encompass a wide area from food processing to food packaging. In food processing, theses nanostructures can be used as food additives, carriers for smart delivery of nutrients, anti-caking agents, antimicrobial agents, fillers for improving mechanical strength and durability of the packaging material, etc. whereas food nanosensing can be applied to achieve better food quality and safety evaluation [25].

Nanotechnology in Food Industry

The nanotechnology offers a variety of applications in food industry. They include improved taste, texture, consistency and shelf-life.

Nano-encapsulations mask tastes, control interactions of active ingredients with the food matrix, control the release of the active agents, ensure availability at a target time and specific rate, and protect them from moisture, heat [26], chemical, or biological degradation. Moreover, these materials possess ability to penetrate into tissues due to their smaller size and thus allow efficient delivery of active compounds to target sites in the body [27].

Texture, Taste, and Appearance of Food items

In order to improve the texture and taste of the food items nanoencapsulation techniques are normally used by the people. For encapsulating cyanidin-3-O-glucoside (C3G) molecules along with soybean seed H-2 subunit ferritin (rH-2) are used.

Nutritional Value

A majority of bioactive compounds such as lipids, proteins, carbohydrates, and vitamins are sensitive to high acidic environment. Encapsulation of these bioactive compounds not only enables them to resist such adverse conditions but also allows them to assimilate readily in food products, which is quite hard to achieve in non-capsulated form due to low water-solubility of these bioactive compounds. Nanoparticles-based tiny edible capsules with the aim to improve delivery of medicines, vitamins or fragile micronutrients in the daily foods are being created to provide significant health benefits [28, 29].

Shelf-Life

In functional foods where bioactive component often gets degraded and eventually led to inactivation due to the hostile environment, nanoencapsulation of these bioactive components extends the shelf-life of food products by slowing down the degradation processes or prevents degradation until the product is delivered at the target site.

Over the past few years, nanotechnology has revolutionized the food sector. It is observed that these materials have unique properties such as high surface to volume ratio and many other interesting properties such as color, solubility, strength, diffusivity, toxicity, magnetic, optical, thermodynamic, etc. [23, 24]. The rising consumer awareness about food quality and health benefits are impelling the researchers to find the way that can enhance food quality while disturbing least the nutritional value of the product.

7.11 Nanotechnology in Building and Construction Industry

Nano-enabled products are being developed for wide-ranging applications in the construction industry. A few of them are listed below

- Reinforcement of concrete with nanoparticles
- Self-cleaning glasses
- Self-cleaning and anti-pollutant concrete
- Nano-paints
- Nano-enabled wood
- Intelligent sensors to enhance the safety of building against natural and other calamities. Fire-resistant coatings
- UV / IR reflecting Windows

Reinforcement of Concrete with Nanoparticles

The use of nano-silica in concrete increases the mechanical strength of an RCC structure. Nano- silica also decreases water penetration into the concrete, leading to better life and durability. RCC with carbon nanotubes also enhances the strength and wear resistance. Similarly, nanocomposite steel is also having three times higher strength than conventional Steel.

Self-cleaning glasses

In glasses, the self-cleaning property comes from coatings with nanoscale thicknesses that have photocatalytic and hydrophilic (water-attracting) properties. When the surface is subjected to rain or simple washing, hydrophilic action takes place causing water to form into thin, flat sheets (without large droplets). In hydrophilic action, water droplets hitting the surface of the coated glass attract each other, thus forming thin sheets. Surface drainage is enhanced. The water sheet carries off the loosened dirt particles.

Fig. 7.11 The self-cleaning glass (Image: *Pilkington*)

Nanopaints

Paints that serve protective or decorative purposes are ages old. Today nanopaints come in many forms and have been developed with many different needs in mind. Common objectives include improved hardness, glossiness, and color steadfastness. Self-cleaning, anti-staining, anti-scratching anti-bacterial and anti-microbial actions are also common objectives. Many paints are now designed to have self-cleaning actions, wherein dirt and other foreign substances are easily washed away naturally or made to lose enough to be easily removed. Both organic and inorganic nanoparticles find use in relation to pigments. Dispersed inorganic or metallic nanoparticles add hardness and strength. Well-dispersed inorganic nanoparticles can also impart highly sought scratch-resistant qualities to paint. The addition of nanoclays can help in making paints more flame-spread retardant.

Self-cleaning and antipollutant concrete

A highly interesting application for architects is the exploration of self-cleaning concretes. In the case of concrete, the photocatalytic nanoparticles are directly mixed into the concrete or mortar mix at their surfaces. In this process, titanium dioxide particles (anatase) provide photocatalytic effect that causes dirt and other organic matter on the surface to oxidize and degrade. Hydrophilic action from rains or washings cause the impinging water to form a thin-sheet that carry away dirt particles. In addition to self-cleaning properties, the concrete reportedly has capabilities for reducing pollutants such as nitric oxides in the air.

Fig. 7.12 The Jubilee Church in Rome is coated in self-cleaning cement. (Image: italcementi)

Antifogging, Antireflection Characteristics

Many antifogging applications depend on the hydrophilic action of surfaces; the resulting spreading of water droplets becomes largely invisible. Nanoproducts with antireflection characteristics are widely used in architecture as well as in product design to reduce troublesome reflections.

Nano-enabled wood

Impregnation of ZnO_2, or Ag, or TiO_2 nanoparticles in wood can help to reduce their bio-deterioration. This could be a potential substitute for the current practice of using a toxic chemical treatment to prevent the bio-deterioration of wood.

Fig. 7.13 Super-hydrophobic surface nanocoatings that prevent the biodeterioration of wood

One of the important applications of nanotechnology in civil engineering could be the integration of sensors into on-going structure to monitor the safety of the construction. This could result in the early assessment of the damage after a calamity or even after terrorist sabotage [8]. Sensors can also be used to sniff toxic gases and for the early detection of wood-destroying termites.

References

1. Shirsath CA and Shaikh MA, The vital role and applications of nanoelectronics in the Nanotechnology Domain, Int. Res. J. of Science & Engineering, Special Issue A1:167-170, 2017.

2. Glenn Clayton, Nanoscience and Nanotechnology, published by ED-Tech Press, 2018

3. P. Jayarangarao, M.L.N. Acharyulu, Aruna Kumari Nakkella, Nanotechnology a revival approach for scientific development material science, Journal of Information and Computational Science, Volume 10, Issue 3, 2020.

4. Ji-Hyeon Song, Soo-Hong Min, Seung-Gi Kim, Younggyun Cho, Sung-Hoon Ahn, Multi-functionalization Strategies Using Nanomaterials: A Review and Case Study in Sensing Applications, International Journal of Precision Engineering and Manufacturing Green Technology, 2021.

5. Gao, Wei, et al. "Fully integrated wearable sensor arrays for multiplexed in situ perspiration analysis." Nature 529.7587 (2016): 509-514.

6. Sustainable Use of Nanomaterials in Textiles and their Environmental Impact Haleem a Saleem and Syed Javaid Zaidi Materials, Volume 13, 2020.

7. https://www.azonano.com/article.aspx?ArticleID=3058

8. B. S. Murty, P. Shankar, Baldev Raj, B. Brath, James Murday, Textbook of Nanoscience and Nanotechnology, Universities Press, 2013.

9. Deepika, Nanotechnology implications for high performance lubricants, SN Applied Sciences, Volume 2, 2020.

10. Prachi, Pranjali Gautam, Deepa Madathil, A. N. Brijesh Nair, Nanotechnology in Waste Water Treatment: A Review, International Journal of ChemTech Research, Volume 5, pp2303-2308, 2013.

11. Ali Mansoori, G., et al. "Environmental application of nanotechnology." Annual review of nano research (2008): 439-493.

12. Ankit Nagar and Thalappil Pradeep, Clean Water through Nanotechnology: Needs, Gaps, and Fulfillment, ACS Nano, Volume 14, pp 6420−6435, 2020.

13. https://www.iinano.org/energy

14. Yan-Gang Bi, Jing Feng*, Jin-Hai Ji, Fang-Shun Yi, Yun-Fei Li, Yue-Feng Liu, Xu-LinZhang and Hong-Bo Sun, Nanostructures induced light harvesting enhancement in organic photovoltaics, Nanophotonics 2018; 7(2): 371–391.

15. Rakesh Rathi, Nanotechnology, S. Chand Publishing, 2009.

16. Anna Pratima Nikalje, Nanotechnology and its Applications in Medicine, Medicinal chemistry, volume5(2), 2015.

17. hangW, WangY, LeeBT, LiuC, WeiG, etal., A novel nanoscale dispersed eye ointment for the treatment of dry eye disease, Nanotechnology, Volume 25, 2014.

18. Ansujap. Mathew, Robin augustine, Nandakumar kalarikkal, and Sabu Thomas, Nanomedicine and Tissue Engineering, Apple academic press, 2016.

19. Logothetidis S, Nanotechnology in Medicine: The Medicine of Tomorrow and Nanomedicine, Hippokratia, Volume 10 (1), pp: 7-21, 2006.

20. Ngo, Christian, van de Voorde, Marcel, Nanotechnology in a Nutshell, Atlant is press, 2014.

21. Alaa Y. Ghidan and Tawfiq M. Al Antary, Applications of Nanotechnology in Agriculture, Applications of Nano biotechnology, Intech open book series, 2019.

22. Pramanik, Pragati, et al. "Application of nanotechnology in agriculture." Environmental Nanotechnology Volume 4. Springer, Cham, 2020. 317-348.

23. Rai M, Yadav A, Gade A. (2009). Silver nanoparticles as a new generation of antimicrobials. Biotechnol Adv 27:76–83

24. Gupta, A., Eral, H. B., Hatton, T. A., and Doyle, P. S. (2016). Nano emulsions: formation, properties and applications. Soft Matter 12, 2826–2841. doi: 10.1039/c5sm02958a

25. Ezhilarasi, P. N., et al. "Nanoencapsulation techniques for food bioactive components: a review." Food and Bioprocess Technology 6.3 (2013): 628-647.

26. Ubbink, Job, and Jessica Krüger. "Physical approaches for the delivery of active ingredients in foods." Trends in Food Science & Technology 17.5 (2006): 244-254.

27. Lamprecht, Alf, et al. "Lipid nanocarriers as drug delivery system for ibuprofen in pain treatment." International journal of pharmaceutics 278.2 (2004): 407-414.

28. Yan, S. S., and Gilbert, J. M. (2004). Antimicrobial drug delivery in food animals and microbial food safety concerns: an overview of in vitro and in vivo factors potentially affecting the animal gut microflora. Adv. Drug Deliv. Rev. 56, 1497–1521.

29. Koo, O. M., Rubinstein, I., and Onyuksel, H. (2005). Role of nanotechnology in targeted drug delivery and imaging: a concise review. Nanomed. Nanotechnol. Biol. Med. 1, 193–212.

Chapter 8

Nanoscale Devices

8.1 Introduction

Over the last decade the interest in nanoscale materials and their applications in variety of new and novel devices has been increasing tremendously. The unique properties of nanoscale materials and their outstanding performance especially when they are used in nanoscale devices are responsible for the emergence of nanoscale devices. The fascinating and interesting properties of nanoscale devices opened new and sometimes unexpected fields of applications. Today, the extensive applications vary from the detection of explosives, drugs and fissionable materials to bio- and infrared-sensors, spintronic devices, data storage media, magnetic read heads for computer hard disks, single-electron devices, microwave electronic devices, and many more [1].

Nanoscale devices are the ones that are ten to fifteen thousand times smaller than human cells. These devices can manipulate matter on an atomic scale without any difficult. Examples of nanoscale devices are synthetic molecular motors such as high electron mobility transistors, graphene-based transistors, resonant tunneling diodes etc., [2].

Nanoscale phenomenon is responsible for several new and interesting phenomenon including quantum confinement, single-electron effects in electronics, near-field behavior in optics and electromagnetics, single-domain effects in magnetics etc., Some other effects including mechanical, fluidic, and biological systems. Both fundamental materials processing, and device technologies, as well as the integration of such technologies into complex systems with consideration of system drivers and constraints as guides for the development of new materials and devices, are encompassed within this program.

8.2 Application of Nanoscale Devices for Photovoltaics

Solar cell technology directly converts the clean, abundant energy from the sun into electricity. The cost and inefficiency of these solar panels have prevented them from becoming an economically competitive form of everyday power generation. However, with the advent of nanoscience and nanotechnology new ideas and new hopes helped the people working in this area have started looking up to develop cost effective systems. In recent times, it has been proved that the nanomaterials such as nanowires, nanotubes and quantum dots are very promising candidates, due to their novel thermal, electrical, and optical properties [3]. It was also proved that silicon nanowire, carbon nanotube and semiconductor quantum dots are also the potential candidates due to their high optical absorption and low electron-phonon coupling.

It was concluded based on the data available that the vertically aligned multiwalled carbon nanotube arrays are nearly perfect absorber in the visible spectrum. Silicon nanowire arrays are less absorptive than carbon nanotube, but we propose and demonstrate that their optical absorption can be greatly enhanced by introducing structural randomness, including random positioning, diameter and length. Later on it was shown that the overall efficiency of nanotube/nanowire array solar cells are also very high indicating that they can also be exploited. Thus one may conclude that nanoscale materials may be exploited for achieving more efficient conversion of solar energy for day to-day applications.

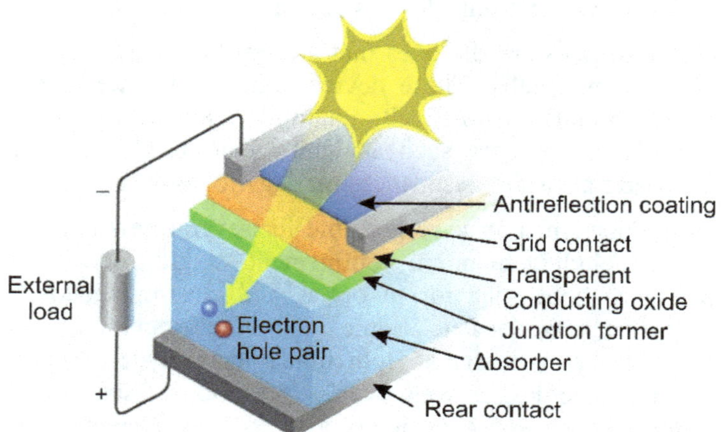

Fig. 8.1 Schematic of a photovoltaic cell. rs. Image Credit: NREL

8.3 Application of Nanoscale Devices for Medical Diagnostics

The real-time, personalized and highly sensitive early-stage diagnosis of disease remains an important challenge in modern medicine. In fact, a successful screening and diagnosis of a particular disease depends too heavily on the skill and experience of the radiological practitioner.

However, with the ability to interact with matter at the nanoscale and the development of nanotechnology architectures and materials could provide excellent techniques beyond the limits of conventional diagnostic modalities [4].

Nanomaterials have a major role in medical diagnostics, including efficient targeted drug delivery, development of artificial cells, smart drugs, immune isolation therapies etc., The other areas where nanotechnology is having some potential include chromosome replacement, cell surgery etc. Another area where a lot of work is going on is in the replacement of defective or incorrectly functioning cells and organs, especially related to metabolic functions. Nanotechnologies promise to extend the limits of current molecular diagnostics and enable point-of-care diagnosis, integration of diagnostics with therapeutics, and development of personalized medicine [5]. The most important clinical applications of currently available nanotechnology are in the areas of DNA detection, biomarker discovery, cancer diagnosis and detection of infectious microorganisms. Nanomedicine promises to play an important role in the future development of diagnostic and therapeutic methods.

8.4   Nanoscale Devices for Electronic Applications

Nanoelectronics in a very short time from now may lead to the new industrial revolution. This technology deals with the characterization, manipulation and fabrication of electronic devices at the nanoscale level with enhanced capabilities, reduced weight and low power consumption. Electronic devices from computers to smart cell phones have become part and parcel of our life and modern living is incomplete without these gadgets. However, increasing thermal issues and the manufacturing costs associated with traditional semiconductor technology have hindered further development in device technology. Nevertheless, nanoscale electronic devices are very small devices to overcome limits on scalability, which provide alternative options in terms of ease of processing, better flexibility, low cost, significant increase in speed processing capability. Currently, for commercial use, devices with featured sizes of 14 nm or below can be fabricated and new computer microprocessors have less than 50 nm of feature size.

The superior electronic properties of materials come when electrons are confined to structures that are smaller than mean free path of electrons in normal solids. In the nanotechnology era, a variety of nanomaterials have been synthesized, which show unique physical, mechanical, electrical, electronic, and photonic properties. These nanomaterials have been used as functional elements in device applications. For example, carbon nanotubes, semiconductor nanowires, quantum dots, graphene, graphene oxide, and transition metal dichalcogenides have been used to replace rectifiers, junction transistors, field effect transistors, CMOS, RAM, and a number of other silicon-based devices [6]. It is believed that nanoelectronic science will develop new nanoscale circuits, processors, and means of storing or transferring information, which can offer greater versatility because of faster data transfer, more on-the-go processing capabilities, and larger data memories. [7].

References

1. Rudolf Gross, AnatolieSidorenko, LenarTagirov, Nanoscale Devices, Fundamentals andApplications, Springer, NATO Science Series II: Mathematics, Physics and Chemistry,2006, vol-233, XXII, p.378.

2. https://www.nature.com/subjects/nanoscale-devices.

3. Kehan Yu and Junhong Chen, Enhancing Solar Cell Efficiencies through 1-DNanostructures, Nanoscale Research Letters, 2009, Vol-4, p.1-10.

4. Ye Hu, Daniel H Fine, EnnioTasciotti, Ali Bouamrani, Mauro Ferrari, Nanodevices indiagnostics, Wiley interdisciplinary reviews. Nanomedicine and nonobiotechnology,2011, Vol-3(1), p.11-32.

5. Michael J. Mitchell et al, Engineering precision nanoparticles for drug delivery, Naturereviews-Drug discovery, 2021, Vol-20.

6. Khurshed Ahmed Shah, Farooq Ahmed Khanday, Nanoscale Electronic Devices and theirApplications, First edition, CRC Press, 2021.

7. Anuhya and Grace Eunice, Potential Impact of Nanomaterials in Information and Communication Technologies, Indian Journal of Research in Pharmacy and Biotechnology, 2016, Vol-4(6), p.267-270.

Chapter 9

Concerns and Challenges of Nanotechnology

9.1 Introduction

Although there are several applications of nanotechnology, there are several risks also,. Some of the risks might influence human health, safety issues; transitional effects such as displacement of traditional industries as the products of nanotechnology become dominant. These may be particularly important if potential negative effects of nanoparticles are overlooked.

9.2 Environment Pollution

Environmental pollution is the key concern among the public. Concern of the common public for the adverse impacts is to be educated. Nanoparticles might enter the environment in a number of ways. They include, accidental escape during manufacturing, disposal of waste products and recycling process that involve grinding, abrasion or heat.

The environmental issue of Nanotechnology is an important area. Although, only a few scientific studies available on the impact of nanoparticles on the environment, one may conclude that some properties of nanoparticles are definite risk to the environment [1].

9.3 Toxicity

Employees who use nanomaterials, both in research or in production may be exposed due to inhalation, dermal contact, or ingestion, depending upon how employees handle them. Although the potential health effects of such exposures are not fully understood at this point of time, scientific studies indicate that at least some of these materials are biologically active and may penetrate intact human skin, and even lungs may be exposed [2]. The extremely small size of nanomaterials means that they are much more readily taken up by the human body than larger sized particles. How these nanoparticles behave inside the organism is one of the issues that need to be resolved.

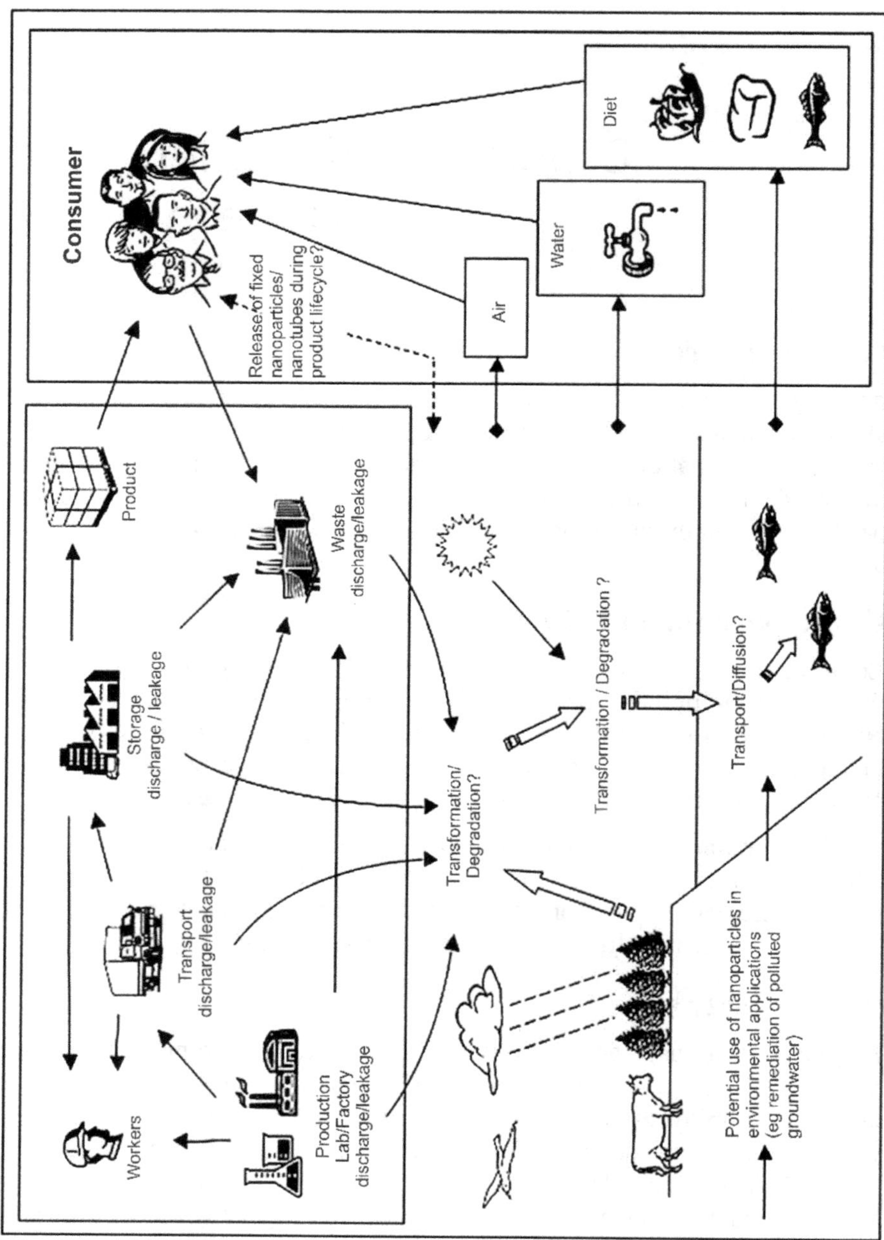

Fig. 9.1 Possible exposure routes of nanoparticles with atmosphere (Image adopted from 2004, The Royal Society & the Royal Academy of Engineering report on Nano science and nanotechnologies)

9.4   Privacy

In food packaging, obligatory surveillance of food during production and distribution with RFID tags is essential to protect consumer health. However, some customers feel that their privacy is taken by the technology. In fact, the RFID nanosensor continues to monitor and save data of the all the users. This data can be recovered by unauthorized people and their privacy is lost.

9.5   Concerns of Nanosensors

Although several are people using nanosensors extensively for various purposes, there are many unanswered questions regarding their toxicology issues which may limit their application in biological systems. In fact, a few nanosensors may even affect the cell metabolism changing the cellular and molecular profiles and making it difficult to work with them [3]. The main problem with using nanosensors especially in food may be due to the fact that it effects the overall toxicity. At present, it is not clear how the nanosensors affect the soil, plants and humans in the long-term. Moreover, it is difficult to address the problem of nanoparticle toxicity as it depends on several factors. They include, the type, size and dosage of the particle as well as environmental variables including pH, temperature and humidity.

9.6   Concerns in Military

Although there are many benefits nanotechnology to military both on the battlefield and in civilian life, there are many other related issues to be discussed at length. As this sort of technology starts to expand back into the civilian realm, there will be an enormous challenge to regulate the use of nanotechnology. Widespread availability of these devices would inevitably may be misused by the criminals and terrorists. Nanotechnology could be used for developing atomic weapons. In fact, this technology may be used to create and develop novel and improved weapons. Atomic weapons could be more accessible and destructive. Another interesting thing developed is "smart bullet," a computerized bullet that could be controlled and aimed at the target very accurately. These developments may prove a boon for the military.

9.7   Concerns in Nanotextiles

People are developing variety of techniques for the production of nano-textiles leading to the enhanced manufacturing cost. This may affect the quality control, homogeneity of nanoparticle dispersion through the fabric. However, there are some implications of nanoparticle exposure, for human health and for the environment. Although recent research work has focused on this issue, total understanding of the toxicological effects of the various kinds of nanoparticles is yet to understand [4].

Disadvantages

➢ Expensive as its production cost is high.

➢ Some of the existing production industries could be effected in adverse.

➢ Impact of nanotechnology on the environment is still unknown

9.8   Challenges of Nanotechnology

Because of lack of understanding of the potential and risks of nanotechnology, society is in total danger. In this connection, several people who are deeply involved in this field cautioned the society about various types of risks involved and suggested ways and means for "Safe handling of Nanotechnology" and identified five grand challenges. People suggested that research in nanotechnology should "stimulate research that is imaginative, innovative, timely and relevant to the safety of Nanotechnology [5].   Some of the suggestions are,

1. Instruments to assess environmental exposure to nanomaterials
2. Methods to evaluate the toxicity of nanomaterials
3. Models for predicting the potential impact of new engineered nanomaterials
4. Ways of evaluating the impact of nanomaterials across their life cycle, and
5. Strategic programs to enable risk-focused research.

References

1. Environmental Science &Technology- 2006, 40, 1401- 1407
2. J. Environ. Sci. Health C- Environ Carcino Ecotoxicol Rev.- 2009, 27, 1-35.
3. Nature Nanotechnology – 2021, 16, 251-265
4. IOP Conference series Materials Science & Engg.- 2017, 254, 102002
5. Polymer Surface & Interface- 2016, 48, 371- 389.